STEAM AGE

JULIAN HOLLAND

D&C

David and Charles

CONTENTS

INTRODUCTION

What is steam? According to the *Concise Oxford Dictionary*, it is 'the gas into which water is changed by boiling, used as a source of power by virtue of its expansion of volume'. Well, there you have the answer in all its simplicity. And it was the application of this concept by a few inventive geniuses, most of them British, that changed the face of the world in the 18th and 19th centuries.

The power of steam, however, had been recognised for centuries – the first known example, a ball spun by jets of steam, was described by Hero of Alexandria in the 1st century AD. By the 17th century various ideas had been put forward for steam pumps and turbines but it wasn't until the end of that century that the Devonshire engineer, Thomas Savery, patented an atmospheric steam engine to pump water out of mines.

In reality this invention wasn't very successful, having no moving parts apart from a few taps and being unable to pump water from a depth of more than 30 feet. However the first major leap forward in steam technology came in 1710 when Cornish

Baptist preacher and engineer, Thomas Newcomen, built the world's first practical steam pumping engine. Known as the beam engine, this was the first steam engine to employ a cylinder and moveable piston and models were soon being used to drain deep mines throughout Britain and Europe.

By the late 18th century the inventive mind of Scottish engineer James Watt had further refined the steam pumping engine and his successful partnership with Birmingham foundry owner Matthew Boulton played a major role in kick-starting Britain's Industrial Revolution. The combination of abundant coal reserves, inventive minds, cheap labour and entrepreneurial spirit made Britain the world leader in steam technology. It wasn't long before this technology was being further developed and used across a wide variety of industrial and transport applications.

From the early 19th century the evolution of the steam engine in Britain stepped up a gear to herald a series of world firsts – Richard Trevithick's *Penydarren* steam railway locomotive of 1802, George Stephenson's Stockton & Darlington Railway of 1825 (the world's first public railway) and his Liverpool & Manchester Railway of 1830 (the world's first inter-city railway). The development of the steam railway locomotive continued unabated until the 1930s, culminating in Nigel Gresley's iconic streamlined locomotives of which *Mallard* still holds the world rail speed record for steam of 126mph.

Britain also led the way with the evolution of steamships – from William Symington's *Charlotte Dundas* of 1803 (the world's first successful steam-powered boat) and Francis Petit Smith's SS *Archimedes* (the world's first screw-driven steamship) to HMS *Rattler* (the world's first steam screw-driven warship) and Isambard Kingdom Brunel's giant transatlantic steam-driven liners of which the *Great Eastern* was by far the largest. British world firsts continued unabated, culminating in Charles Parsons' steam turbine ship *Turbinia* that was so vividly demonstrated to an unsuspecting Royal Navy in 1897. By the early 20th century Britain was undoubtedly ruling the waves with its new high-speed 'Dreadnought' battleships and transatlantic liners.

By the end of the 19th century steam had become the prime mover in Britain's race to become the dominant world power, not only at sea but also at home where it drove the country's industries and powered its transport system. While the extensive railway network was the key to this success, steam also saw more short-lived applications for road transport and agricultural machinery – steam-powered carriages, omnibuses, trams, wagons, lorries, road rollers, mechanical shovels, ploughs, tractors and threshing machines all played their part. However by the 1930s the onslaught of the internal combustion engine and punitive road taxes for steam road vehicles soon put an end to all that. Last but not least we must also remember some very British quirky inventions such as the flying steam carriage of 1842 that never really got off the ground and the steam powered submarine which actually saw service at the end of World War 1.

While steam in the factories and on the railways and roads is now a fading memory, it should be worth noting that much of our electricity is, thanks to inventors such as Charles Parsons, still generated by steam turbines in our coal, gas and nuclear power stations. Even modern nuclear powered submarines use steam turbines to drive their propulsion units.

Author's footnote: In Britain we are extremely lucky to have the most amazing steam preservation movement. Lovingly restored and cared for by happy bands of mainly volunteer enthusiasts, steam engines can still be seen at work on our waterways, roads and railways and at hundreds of steam events around the country. Britain's world-beating steam heritage is still very much alive!

First Faltering Steps
Thomas Savery's steam pump

Born in the mid-17th century in Modbury, Devon, Thomas Savery became a military engineer and inventor. Turning his mind to the problem of pumping water from within the Cornish mines, in 1698 he patented an atmospheric steam engine which he demonstrated to the Royal Society a year later. Savery's invention closely followed an idea for such a machine that had been published by Edward Somerset, 2nd Marquess of Worcester, as early as 1662.

Savery's early steam engine had no working parts apart from some taps. Steam was first raised in a boiler and was then allowed to blow into a large enclosed reservoir and down a pipe to the water at the bottom of the mine. When the system was full of hot steam the tap between the boiler and the reservoir was closed (a tap to the outlet pipe at the top of the reservoir was also kept closed) and as the steam cooled and condensed (thus creating a partial vacuum) atmospheric pressure forced the water at the bottom of the mine up the pipe and into the reservoir. Once this was full the tap at the bottom of the reservoir was closed and the tap to the outlet pipe at the top was opened. More steam was raised in the boiler, the tap to the reservoir was opened, and the water in the reservoir was forced by steam pressure up the outlet pipe and out of the mine.

In theory, Savery's steam pump was a godsend to owners of mines prone to flooding but in practice it had many faults including being unable to raise water from more than 30 feet. Deep mines therefore needed a whole series of Savery's pumps working together at different levels. The high-

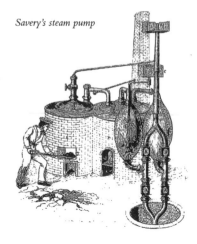

Savery's steam pump

pressure steam required to force water to the surface also created problems with the soldered joints that required constant repair. Despite this, several of his pumps were successfully used for some years to maintain water supplies at various locations in London including Hampton Court.

SAVERY'S FINANCIAL GAIN

Savery's original patent was extended in 1699 by an Act of Parliament known as the Fire Engine Act. Fortunately for him, his patent covered all types of steam engines that pumped water and when the Devonshire engineer Thomas Newcomen (see 'A Leap in Technology') created the world's first practical steam pumping engine in the early 18th century he was forced to go into partnership with Savery. Although Savery died in 1715, royalties accrued from Newcomen's engines continued to be paid into his estate until the patent expired in 1733.

A Leap in Technology
Thomas Newcomen's atmospheric steam pumps

Thomas Newcomen was born in Dartmouth, Devon, in 1664. He became a Baptist preacher, engineer and businessman and is credited with designing the world's first practical steam pumping engine. Living in south-west England where he was no doubt influenced by the constant flooding of Cornish tin mines, Newcomen improved Thomas Savery's steam pump (see 'First Faltering Steps') by replacing Savery's steam reservoir with a cylinder and moveable piston.

Originally invented by the French inventor Denis Papin in 1690 (who is also credited with inventing the pressure cooker), the piston was forced up inside the cylinder by the pressure of steam from a boiler and sucked down by the vacuum created by water-cooled condensed steam. The top end of the moving piston (sealed by an air-tight flap at the top of the cylinder) was attached by a chain to one end of a rocking wooden beam; the other end of the beam was similarly attached to a pump at the bottom of the mine. The up-and-down movement of the beam caused water at the bottom of the mine to be drawn into the pump and forced to the surface.

Introduced in 1710, Newcomen's beam engines were a major leap forward in technology and were soon being used to drain deep mines throughout Britain and in Europe. Water could now be raised from a depth of 150 feet and the first example replaced a team of horses that had previously been used to operate the pumps.

Despite his world-beating invention Newcomen failed to become a wealthy man – royalties for his beam engines went to Savery whose 1699 patent covered 'all types of steam engines that pumped water' and, sadly, Newcomen died in 1729, four years before the patent expired. Despite major technological developments of the steam engine made later by James Watt and Matthew Boulton (see 'A Steam Partnership'), Newcomen's 'Common Engines', as they were called, were built in large numbers with some

remaining in commercial use for well over 150 years.

One of his early beam engines can be seen at the Newcomen Engine House in Dartmouth which is open to the public as a memorial to one of the forefathers of the Industrial Revolution. Newcomen engines can also be seen at the Science Museum in London, the Elsecar Heritage Centre in South Yorkshire and the Henry Ford Museum in Dearborn, Michigan, USA.

THE HORNBLOWER FAMILY

This Cornish family were pioneers of steam power in the 18th and early 19th centuries.

JOSEPH HORNBLOWER (1696-1762)

Worked as an installer of Newcomen beam engines in Cornish mines from the 1720s to 1740.

JONATHAN HORNBLOWER (1717-1780)

Eldest son of Joseph Hornblower. Learnt his trade from his father as a builder and installer of Newcomen beam engines in Cornish tin mines.

JOSIAH HORNBLOWER (1729-1809)

Youngest son of Joseph Hornblower.

Following his brother, Jonathan, he became an installer of Newcomen engines in Cornwall and an expert on mining. He travelled to North America in the 1750s to advise on copper mining and built, using parts brought over from Britain, what some claim to be the first steam engine on the American continent. A mechanical and civil engineer, Josiah also dabbled in politics and later served as a judge.

JABEZ CARTER HORNBLOWER (1744-1814) *The eldest son of Jonathan, Jabez followed in the family tradition as installer of Newcomen beam engines in Cornwall before working for Boulton & Watt (see 'A Steam Partnership') as a builder of the new Watt steam engines. He later set up business with John Maberley and built a new steam engine; however parts of it infringed James Watt's patent and he was successfully sued. His business collapsed and Jabez ended up in a debtors' prison.*

JONATHAN HORNBLOWER JNR (1753-1815) *The youngest son of Jonathan Snr, he invented a compound steam engine in 1781,*

but his work on this was successfully stopped by James Watt (see 'A Steam Partnership') who claimed that it was his intellectual property. The compound engine was later revived successfully by Arthur Woolf (see 'Improving Efficiency'). In the meantime, Jonathan Jnr went on to become a successful steam engineer and died a wealthy man.

A Steam Partnership
James Watt and Matthew Boulton

J ames Watt was born in 1736 in Greenock, Scotland. In 1754 he travelled to London where he studied instrument making before returning to Glasgow where he made and repaired navigational measuring instruments, scales, barometers and telescopes. Unable to set up his own business because he had not completed the statutory period as an apprentice, Watt went on to repair astronomical instruments at Glasgow University until in 1759 he joined forces with architect and businessman John Craig to open a workshop manufacturing measuring and musical instruments and toys.

While in his workshop Watt started experimenting with and learning about steam engines, in 1763 managing to get his hands on a broken model Newcomen beam engine. While repairing it he discovered the engine's inadequacies, notably its inefficient use of steam when condensed in the cylinder by injected cold water. Watt set about building a small-scale steam engine with a separate condenser and in 1765 demonstrated its vastly improved capabilities. In order to fund the building of a full-size prototype Watt went into partnership with John Roebuck of the Carron Iron Works but, despite obtaining a patent for his invention, Watt's dream was shattered when Roebuck became bankrupt. Following this setback, Watt's patent rights were acquired by Matthew Boulton (1728–1809), owner of the Soho Foundry in Birmingham. With Boulton's business acumen and precision engineering capabilities Watt was at last able to produce his first full-size working steam engine.

Incorporating further improvements such as cooling the condenser in a tank of cold water,

the first production model of the new Boulton & Watt steam engine was completed in 1775. Using around 75 per cent less coal than the Newcomen engine, it became an instant success and models were soon in use pumping water from deep mines around the country. The engines, weighing many tons, were from 1795 totally prefabricated in Birmingham, then reassembled by the company's men at the mines – Boulton & Watt also charged a licence fee to the owners which was based on the fuel saved. Initially used to pump mines, the engines had many other industrial applications and played a major part in the burgeoning growth of Britain's Industrial Revolution. Moreover in 1807 the company supplied the engine for Robert Fulton's *North River Steamboat*, the first steamboat on the Hudson River in North America.

With a full order book from customers around Europe and North America the company continued to develop their engines by introducing double acting pistons, rotative motion, centrifugal governors and parallel motion to connect between piston and beam. By the time of his death in 1819 James Watt had become a wealthy man while the company of Boulton & Watt continued to build steam engines until the end of the 19th century.

Many Boulton & Watt steam engines have survived into the 21st century – in fact the oldest working engine in the world, Boulton & Watt's Smethwick Engine of 1779, is now in working order at Thinktank Science Museum in Birmingham. Dating from 1812, the oldest still to be found in its original engine house can be seen occasionally working at the Crofton Pumping Station on the Kennet & Avon Canal. Other examples can be seen at the Science Museum in London, the Henry Ford Museum in Dearborn, Michigan, USA and the Powerhouse Museum in Sydney, Australia.

A Prolific Inventor
William Murdoch's steam carriage

S cottish engineer and prolific inventor William Murdoch was born in Cumnock, Ayrshire, in 1754. Excelling in maths at school,

Murdoch initially worked for his millwright father on a local estate where he learned the principles of mechanics. Seeking a job with James Watt, he travelled to Birmingham in 1777 and so began his illustrious career with Boulton & Watt. Starting as a pattern maker, by 1779 Murdoch's skills were being put to good use by the company as he was sent to Redruth in Cornwall to oversee the erection, maintenance and repair of engines that had been sold to Cornish tin mines. He became adept at tweaking the Boulton & Watt engines and improving their efficiency, thus generating increased revenues for the company as it was paid a licence fee based on the amount of fuel saved. His other task was to keep an eye on the competition to ensure there were no infringements of company patents.

Murdoch's enquiring mind led to many inventions for improving steam engines while he was based in Cornwall, although as an employee of Boulton & Watt the subsequent patents became the property of the company. Of note was his sun and planet gear that achieved rotary

Murdoch's steam carriage

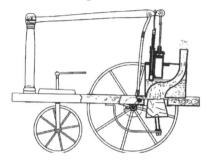

motion by turning a drive shaft – this system became widely used in many industrial applications. Murdoch also invented the pneumatic despatch system that used compressed air to propel a cylinder through a tube – this was widely used to pass messages, receipts or cash within department stores well into the 20th century.

AN ECCENTRIC INVENTOR

In search of employment, William Murdoch walked the 300 miles from his home in Ayshire to Birmingham wearing a wooden hat that he had turned on a lathe. Matthew Boulton was so impressed with the hat that he gave Murdoch a job. In later years, Murdoch was secretly testing his model steam carriage on a Cornish road at night. Unknown to him it was seen by a local vicar who mistook the snorting machine for the devil! Even his house in Birmingham was filled with weird and whacky inventions such as a doorbell that was operated by compressed air.

In 1784, Murdoch built a working model of a steam self-propelled road carriage – while not the first in the world (the French engineer Cugnot had built two full-size working steam vehicles some years earlier) it was a first for Britain. Using high-pressure steam the miniature three-wheeled vehicle successfully ran around the floor of Murdoch's house in Redruth. However his heavy workload with Boulton & Watt and his marriage got in the way of any serious development with his design for some years. A second improved model was eventually made and, in 1795, was successfully demonstrated in a Truro hotel – this was the first public demonstration of a working steam locomotive in Britain. However Murdoch's employers, disliking high-pressure steam, were not supportive of his invention and it was left to his neighbour, a certain Richard Trevithick (see 'The First Road Locomotive'), to develop the steam road vehicle. Murdoch's model steam road carriage can be seen today at the Thinktank Science Museum in Birmingham. A prolific inventor, Murdoch went on to discover iron cement which was used to strengthen the joints in steam engines, to invent gas lighting and to develop the use of steam engines to propel boats. Murdoch eventually took charge of the marine business for Boulton & Watt that built around 50 marine engines in the early 19th century. In his later years he lived near the company's Soho Works in Birmingham – he died in 1839 at the grand old age of 85.

Beam Engine Survivors
Where to see stationary beam engines

BLAGDON LAKE PUMPING STATION
Station Road, Blagdon, North Somerset BS40 7UN
Website: www.bristolwater.co.uk
A reservoir to supply Bristol with clean water was built at Blagdon in the late 19th century. Four Woolf compound rotative beam pumping engines built by Glenfield & Kennedy of Kilmarnock were installed between 1900 and 1905. These were replaced by electric pumps in 1949 although two of the beam engines were retained as museum pieces. One of them is now operated by an electric motor at Blagdon Lake Visitor Centre on Sundays between April and August.

BOLTON STEAM MUSEUM
Atlas Mills, Chorley Old Road, Bolton, Lancashire BL1 4EU
Website: www.nmes.org
Located in what was one of the largest cotton spinning mills in Britain, Bolton Steam Museum contains the largest collection of working steam engines in the country. Originally built to work in Lancashire's mills, the 25 engines in the museum have been restored to working order by the Northern Mill Engine Society. Exhibits include the 1840 Crossfield Beam Engine and the 1870 Cellarsclough MacNaughted Beam Engine, as well as other types of steam engine that kept the Lancashire mills humming through the 19th and early 20th centuries.

CROFTON PUMPING STATION

**Crofton, Marlborough, Wiltshire
SN8 3DW
Tel: 01672 870300
Website: www.
croftonbeamengines.org**

Built in 1807 to pump water to the summit of the Kennet & Avon Canal, Crofton Pumping Station houses two restored working beam engines – an 1812 Boulton & Watt and an 1846 Harvey engine. These are steamed for the public on several weekends during the summer months and while working actually carry out the job for which they were built.

CROSSNESS PUMPING STATION

**The Old Works, Crossness STW,
Belvedere Road, Abbey Wood,
London SE2 9AQ
Tel: 020 8311 3711
Website: www.crossness.org.uk**

Described as a Victorian cathedral of ironwork, and featuring some of the most beautiful ornamental cast ironwork in Britain, Crossness Pumping Station was opened in 1865 as part of the London sewerage system. It houses the largest four remaining rotative beam engines in the world with 47-ton beams

and 52-ton flywheels. One of these engines, 'Prince Consort' has been restored to working order and now runs on open days.

DOGSDYKE ENGINE

**Bridge Farm, Sleaford Road,
Tattershall, Lincolnshire LN4 4JG
Tel: 01636 707642
Website: www.dogdyke.com**

Dogsdyke Steam Pumping Station was built in the mid-19th century to drain farmland in Lincolnshire. The original Bradley & Craven steam beam engine and scoop wheel has been preserved and can be seen working on weekends during the summer months.

EASTNEY BEAM ENGINE HOUSE

**Henderson Road, Eastney,
Portsmouth, Hampshire PO4 9JF
Tel: 023 9282 7261
Website: www.
portsmouthmuseums.co.uk**

Designed to pump sewage from Portsmouth, a pair of original 1887 Boulton & Watt beam engines and pumps are housed in this impressive Victorian building. They were kept in working order until 1954 when they were retired. Since restoration, they can be seen in action on Events

Days during the summer months.

ELSECAR HERITAGE CENTRE
Wath Road, Elsecar, South
Yorkshire S74 8HJ
Tel: 01226 740203
Website: www.elsecar-heritage-
centre.co.uk
Located in the former ironworks
and colliery workshops of Earl
Fitzwilliam, the heritage centre
houses the only Newcomen beam
engine in the world that has remained
in its original location. Installed
in 1795 to pump water out of the
colliery, it continued in operation
until the 1920s. There are now plans
to restore it to working order.

KEW BRIDGE STEAM MUSEUM
Green Dragon Lane, Brentford,
Middlesex TW8 0EN
Tel: 020 8568 4757
Website: www.kbsm.org
Built in 1838 to house pumping
engines for the Grand Junction
Waterworks Company, this cathedral-
like building last witnessed a
pumping engine at work in 1958
when a monster Harvey & Co.
standby engine (with 100in-diameter
cylinders) was finally steamed. A
museum since 1973, Kew Bridge now

houses the world's largest collection
of Cornish beam engines – pride
of place goes to Grand Junction 90
Engine which is 40ft high, has a
cylinder diameter of 90in and weighs
in at 250 tons. Restored to working
order, the 'monster' is steamed for
visitors once a month.

LEVANT MINE & BEAM ENGINE
Trewellard, Pendeen, St Just,
Cornwall TR19 7SX
Tel: 01736 786156
Website: www.nationaltrust.org.uk
Installed in its engine house on a
windy Cornish cliff top in 1835, this
Harvey's beam engine is the only
Cornish beam engine in the world
that is still in steam on its original
mine site. Now owned by the
National Trust, the beam engine was
restored to working order in 1992.

MARKFIELD BEAM ENGINE
Markfield Road, Tottenham,
London N15 4RB
Tel: 01707 873628
Website: www.mbeam.org/
A masterpiece of Victorian
engineering and design, the Markfield
Beam Engine is housed in its original
engine house in the former sewage
treatment works for Tottenham.

Now restored to working order, the massive Wood Bros. beam engine with Watt-patented speed governor and parallel motion linkage, and the Arthur Woolf double-expansion compound system, ran at 16 revolutions per minute operating two plunger pumps that could handle four million gallons of sewage each day! It was originally installed in 1888 and can be seen in action on the second Sunday of each month.

PINCHBECK ENGINE
West Marsh Road, Pinchbeck, Spalding, Lincolnshire PE11 3UW
Tel: 01775 725861
Website: www.wellandidb.org.uk
Now restored but turned for visitors by electricity, this A-frame low-pressure single cylinder condensing beam engine was built by the Butterley Company of Leeds in 1833 to drain the Lincolnshire Fens. It continued in service for the Welland and Deepings Internal Drainage Board until 1952 when it was replaced by electric pumps.

RYHOPE ENGINES MUSEUM
Waterworks Road, Ryhope, Sunderland
Tel: 0191 521 0235
Website: www.ryhopeengines. org.uk
Built in 1868 to supply water to Sunderland, the Ryhope Pumping Station houses the two original beam engines that have been kept in working order and are now steamed on certain Open Days between April and October. The identical double-acting, compound rotative beam engines have a working speed of ten strokes per minute and were built by Hawthorn of Newcastle. The beams are 33ft long and each weighs 22 tons.

SMETHWICK ENGINE
Thinktank Birmingham Science Museum, Millennium Point, Curzon Street, Birmingham B4 7XG
Tel: 0121 202 2222
Website: www.thinktank.ac
The oldest working steam engine in the world, the Smethwick Engine was built by Boulton & Watt in 1779 to pump water to the summit of the Birmingham Canal Old Main Line and remained in service until 1892

when it was retired. It was stored for many years by British Waterways until acquired by Birmingham City Council and displayed in the former Birmingham Science Museum. In recent years it has been restored to working order and can be seen at the new Thinktank Science Museum in the city.

TEES COTTAGE PUMPING STATION
Coniscliffe Road, Darlington,
Co. Durham DL3 8TF
Website: www.
communigate.co.uk/ne/
teescottagepumpingstation

This superb Victorian waterworks houses one of the last beam engines to be built in Britain. The rotative two-cylinder double expanding Woolf compound engine was built by Teasdale Bros. in 1904 and remained in service until 1926. Following replacement by electric pumps, it was kept in working order as a standby until 1968 and is now steamed on Open Days between April and October.

The First Steamboat
William Symington's Charlotte Dundas

Born in 1764 in Leadhills, South Lanarkshire, Scotland, William Symington began his apprenticeship as an engineer in the lead mines of the Lowther Hills. Recognising his potential engineering skills after he built his first steam engine in 1785, his mining company sent Symington to study science at Edinburgh University. He returned to design an improved version of Thomas Newcomen's (see 'A Leap in Technology') and James Watt's (see 'A Steam Partnership') atmospheric engines which he patented in 1787.

Symington was the first steam engine inventor in the world to successfully develop this form of propulsion for boats. His first venture on water came in 1788 when he demonstrated a steam-powered paddle boat on Dalswinston Loch near Dumfries. This first steamboat wasn't completely successful but, undeterred, Symington went on to build a second, larger, paddle boat which was successfully demonstrated

The Charlotte Dundas

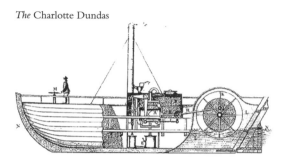

drawing board.

The result was the *Charlotte Dundas* (named after one of Lord Dundas' daughters) which boasted a novel horizontal steam engine that drove a large single rear paddle linked by a crank shaft. At its trials on the newly-opened Forth & Clyde Canal at the end of 1789.

While developing his steamboats Symington also continued to build stationary steam engines for mines in Scotland, and in 1792 he was appointed as consultant steam engineer for the famous Carron Ironworks near Falkirk. In 1801, he patented the world's first horizontal steam engine which he put to good use in his next steamboat.

Recognising the commercial benefits of steamboats for his Forth & Clyde Canal, Thomas (Lord) Dundas provided Symington with financial backing to develop a third boat. Although successfully first tested on the River Carron in 1801 it proved unsuitable for canal work and Lord Dundas sent Symington back to the

in 1803 the steamboat exceeded all expectations by towing two barges, each loaded with 70 tons, along the canal over a distance of 19 miles at an average speed of 2mph. Despite this world-first achievement, the 56ft-long boat was the end of the line for Symington and there was no follow-up order from Lord Dundas. The boat was finally broken up in 1861.

Symington went on to develop his stationary mine engines but, following his installation of a pumping engine at Callender Colliery in Falkirk, he became enmeshed in a long-running legal wrangle that he lost at Edinburgh High Court in 1810. Deeply in debt, Symington moved to London to live with his daughter where he died penniless in 1831. Symington, the designer of the world's first practical steamboat, is buried in St Botulph's, Aldgate, London.

The First Road Locomotive
Richard Trevithick's Puffing Devil

Steam carriage

Born into a Cornish mining family in 1771, Richard Trevithick was soon familiar with the low-pressure steam pumping engines that featured at the deep copper and tin mines in his native county. His own career in mining started in 1790 and by 1797 he had become the engineer at Ding Dong Mine on the Land's End peninsula. It was here that he made modifications to a Boulton & Watt steam pumping engine, incurring the wrath of that company by infringing their patent and avoiding paying royalties to them.

Trevithick had also become acquainted with William Murdoch – at one time they were neighbours in Redruth – and he was particularly interested in the high-pressure model steam carriage that Murdoch had demonstrated to him in 1794.

Trevithick went on to develop the use of high-pressure steam in the final years of the 18th century; his system did away with the Watt-patented condenser, circular motion was transmitted via a crank instead of linear motion via a beam, and he used a double-acting smaller cylinder with improved steam distribution. Consequently, owing to its lighter weight and smaller dimensions, the steam engine was able to carry its own weight plus a loaded carriage.

The big day came in 1801 when, for the first time, Trevithick demonstrated a full-size steam road locomotive in Camborne. Named *Puffing Devil* the carriage, with several passengers on board, rattled through the town much to the locals' apprehension. Despite this iconic moment in the history of world transport, Trevithick's first road carriage had a limited range and soon ran out of steam. A larger steam road carriage was built by Trevithick and launched on an unsuspecting public in London in 1803 – named the *London Steam Carriage* it made a demonstration run through the city but proved uncomfortable and expensive to operate.

The First Clyde Steamer
Henry Bell's Comet

Henry Bell was born into a famous Scottish family of millwrights and civil engineers at Bathgate, West Lothian, in 1767. At the age of 13 years he became an apprentice stonemason, followed by an apprenticeship as a millwright. While he also developed an interest in ship engineering and learnt ship modelling at an early age he was initially unable to pursue his chosen career due, apparently, to his lack of knowledge of the subject!

However Bell was not a man to be deterred – in the first years of the 19th century he was building models of steamboats and even tried to persuade the British Government to build steam-powered ships for the Royal Navy. However his efforts were to no avail, despite the support of the influential Admiral Lord Nelson, so Bell went elsewhere, including America where his ideas were taken more seriously. After exchanging letters with American inventor Robert Fulton, he sent Fulton a model of one of his steamships. This undoubtedly spurred on development of Fulton's first commercial steamboat, the *North River Steamboat,* which began carrying passengers up the Hudson River between New York and Albany in 1807.

Meanwhile back in Glasgow, Bell was continuing to experiment with model steamboats, learning from William Symington's earlier attempt, the *Charlotte Dundas,* and financially supported by his wife, a lady not only in charge of the municipal baths in Helensburgh but one who also found time to run an inn. Finally, in 1811, Bell was able to put into practice everything that he had been working towards as he finished building his first full-size steamboat, the *Comet*.

MAIDEN VOYAGE

Bell's 45ft-long twin paddle steamer **Comet** *was driven by two steam engines and also carried a small sail and a cabin for passengers. Making its maiden voyage in 1812 it was soon carrying fare-paying passengers up and down the Clyde, becoming the first commercially successful steamboat service in Europe. However before long increasing competition from other, more modern, paddle steamers put an end to* **Comet's** *monopoly on the Clyde and she was withdrawn from service, lengthened and re-engined. The rebuilt* **Comet** *re-entered service in 1819 on the Glasgow to Oban and Fort William run (a four-day journey via the newly-opened Crinan Canal) but a year later was shipwrecked near Oban. Undeterred, Bell built the* **Comet II**; *but she was sunk in a collision off Gourock in 1825 with the loss of 62 passengers. This disaster put an end to Bell's entrepreneurial activities and he died in Helensburgh in 1830. He is buried in nearby Row*

churchyard and an obelisk to his memory was erected at Dunglass, overlooking the Clyde. One of the engines from the shipwrecked **Comet** *is on display at the Science Museum in London whilst a replica of the boat is on display in Port Glasgow.*

The First Steam Railway Locomotive
Richard Trevithick and the Penydarren Ironworks

Cornish engineer and inventor Richard Trevithick took out a patent in 1802 for his high-pressure steam engine which was first successfully demonstrated as a stationary engine at Coalbrookdale Ironworks in Shropshire. He also built a high-pressure stationary steam engine to drive a hammer at the Penydarren Ironworks in South Wales. The owner of the works suggested to Trevithick that the engine could be placed on wheels and driven along his horse-drawn cast-iron L-section plateway. The conversion was completed with the drive being transmitted from the single cylinder to the driving wheels via a flywheel and cogs, and the locomotive made its first and only rail journey in 1804, successfully hauling 10 tons of iron and 70 men in five wagons a distance of nearly 10 miles at an average speed of 2mph. This was a world first – the era of steam railway locomotives had begun.

Despite this success the locomotive proved too heavy for the cast-iron plateway and it reverted to its original use as a stationary engine. A second Trevithick steam locomotive was built with flanged wheels for the Wylam Colliery near Newcastle in 1805 but this also proved too heavy for the wooden waggonway on which it ran.

Trevithick's Penydarren locomotive

CATCH-ME-WHO-CAN

In 1808, Trevithick went on to build a circular demonstration steam railway that gave rides to the public in London behind his third locomotive, **Catch Me Who Can.** *Despite its novelty value it wasn't a great success and Trevithick's steam railway locomotive career came to an end, although his designs were later improved by John Blenkinsop (see 'The First Practical Steam Locomotive') for his rack-and-pinion*

Middleton Railway in Leeds. In Trevithick's later years he further developed his stationary high-pressure steam engines for various uses in industry, and inventions such as a steam threshing machine, a steam tug and floating crane, a steam dredger, buoyancy tanks for raising wrecked ships, floating docks and buoys. In 1811, whilst in partnership with a London merchant named Robert Dickinson, Trevithick was declared bankrupt but was subsequently discharged in 1814. In 1816 he travelled to Peru where he became a consultant engineer for silver and copper mines and explored methods of transporting ore back to England. After travelling to Ecuador and Peru, Trevithick ended up in Costa Rica where he considered building a steam railway to connect mines with ports. All of this came to nothing and, virtually penniless, he met Robert Stephenson (see 'Like Father, Like Son') in 1827 who paid for his sea journey back to England. His final years in England were spent as a poverty-stricken inventor and, sadly for the creator of the world's first working steam railway locomotive, he died penniless in Dartford, Kent, in 1833.

Catch Me Who Can

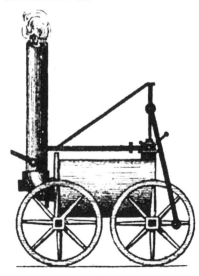

The First Practical Steam Locomotive
John Blenkinsop's rack-and-pinion locomotives

Born in County Durham in 1783, John Blenkinsop served his apprenticeship in Northumberland coal mines until 1808 when he became agent to the owner of collieries on the Middleton Estate near Leeds. A horse-drawn waggonway had already opened in the mid-18th century to convey coal from the mines to Leeds. Known as the Middleton Railway, this was the first railway in Britain to be authorised by an Act of Parliament and, in 1807, the wooden rails on the railway were replaced by stronger iron edge rails.

Following his appointment to the Middleton Collieries, Blenkinsop lost no time in designing a steam railway locomotive to haul the coal trains. While Trevithick's locomotive for the Wylam Colliery in 1805 provided much inspiration, Blenkinsop wanted to develop a more powerful machine capable of hauling much heavier loads. His subsequent six-wheeled design was a marked improvement on the early Trevithick locomotives, weighing in at five tons with innovative features such as twin double-acting cylinders driving geared cogwheels which engaged with a rack rail.

The first two locomotives were built by Matthew Murray (see 'The Jealous Boulton & Watt') of Holbeck in 1812 and proved an immediate success. More followed in 1813 and 1814 and others were built for collieries in Lancashire and Newcastle. These were the world's first practical steam railway locomotives and they stayed in service until the 1830s, by which time the cog-and-rack system had become obsolete following the introduction of the now familiar rolled iron rail and improved adhesion locomotives. Blenkinsop died in Leeds in 1831 and is buried at Rothwell Parish Church.

THE JEALOUS BOULTON & WATT

Matthew Murray (1765–1826), builder of John Blenkinsop's engines for the Middleton Railway, was too nice a person for his own good.

With a complete absence of jealousy for his competitors he was always ready to give technical information where he knew it would be genuinely appreciated. Even in the case of such a formidable rival company as that of Boulton & Watt, Murray does not appear to have withheld anything. We are told that on the occasion of a visit in 1799 of this firm's representatives, Murdoch and a man named Abraham, Murray admitted his visitors into every part of his factory. Murray even presented Murdoch with a specimen of his forge work, no doubt with a view to exciting his astonishment – Murdoch indeed acknowledged that it was the most beautiful and perfect piece of work he had ever seen. However, this friendly treatment appears to have been wasted on Boulton & Watt. They became so jealous of Matthew Murray's company that they purchased a large plot of land adjoining his Holbeck Works in order to prevent the firm from expanding their premises. There appears to be no doubt that Boulton & Watt bought the land entirely for this purpose as they made no use of any part of it. Apparently it remained disused, except for the deposit of dead dogs and other rubbish, for over 50 years.

From Waggonway to Railway
William Hedley's Puffing Billy

Puffing Billy

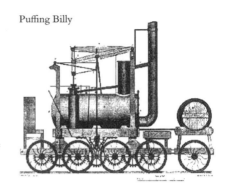

William Hedley was born near Newcastle-upon-Tyne in 1773. At the early age of 22 years he became a colliery manager at Wylam Colliery. At that time the coal from the mine was transported along a five-mile horse-drawn wooden waggonway to Staithes on the River Tyne. Steam haulage was briefly trialled by Hedley in 1805 when Richard Trevithick's second steam railway locomotive was fitted with flanged wheels to run on the wooden rails. However the locomotive was too heavy for the flimsy trackwork and the experiment soon ended.

In 1808 Hedley replaced the wooden rails with cast-iron L-section plate rails and set about designing a steam locomotive with his foreman, Timothy Hackworth (see 'The First Locomotive Superintendent'). The

initial design, based on Trevithick's single-cylinder locomotive, was not a success. Undeterred by this failure the pair built a second engine, again with flangeless wheels to run on the L-section track, but this time incorporating John Blenkinsop's two-cylinder feature as used on the Middleton Railway. There the similarity ended as adhesion was used instead of the cog-and-rack system employed on the latter railway.

As originally built in 1813, Hedley's *Puffing Billy* was a cumbersome eight-ton, eight-wheeled locomotive driven from the pistons by a complicated system of pivoting beams, rods, crankshaft and gears – but it worked, albeit at a maximum speed of 5mph!

The cast-iron plateway continually broke under its weight so Hedley modified the loco to spread the weight more evenly. Along with a sister engine named *Wylam Dilly*, the locomotive continued to operate in this form until 1830 when it was converted into a four-wheeler with flanged wheels to run on newly-laid iron edge rail track. Amazingly these two archaic steam locomotives stayed in service until 1862. The original *Puffing Billy* in its final form can be seen at the Science Museum in London, while *Wylam Dilly* now resides in the Royal Museum in Edinburgh. They are the world's oldest surviving steam locomotives.

William Hedley died in 1843.

Improving Efficiency
Arthur Woolf's compound engines

Cornish engineer Arthur Woolf was born in Camborne in 1766. At the age of 19 years he went to work for the prolific inventor Joseph Bramah at his engineering works in London – Bramah's company was famous for its manufacture of locks, machine tools and hydraulic presses.

After learning his skills as a precision engineer and engine builder, Woolf obtained a patent in 1803 for an improved boiler for producing high-pressure steam. He returned to Cornwall in 1812 where he erected beam engines and built steam stamps for crushing ore, finally becoming Chief Engineer at Harvey & Co. of Hayle in 1816. By then the company had become the leading manufacturer of Cornish beam engines which were exported around the world to drain mines and flooded land. Six of these enormous machines, each incorporating eight beams, were used by the Dutch government to pump the Haarlem Lake dry. Reckoned to be the largest beam engine ever built with a piston diameter of 12ft, one of these machines can today be seen at the Museum de Cruquius in Holland. A man much in demand, Woolf was also engineer at Consolidated Mines in Cornwall from 1818 until 1830.

However, Arthur Woolf's greatest innovation was his compound steam engine which he had patented in 1805 and which vastly improved the efficiency of Cornish beam engines. Designed to extract the maximum

Abbey Pumping Station

amount of energy from steam, the compound engine consists of a high-pressure cylinder in which steam is first expanded then, as it loses heat and pressure, it is exhausted into a larger, low-pressure cylinder. A further development of this arrangement is known as a multiple expansion engine (see 'Marine Triple Expansion Engines') where a whole series of progressively lower pressure, and larger, cylinders are employed. These efficient and smoother machines soon found applications not only in Cornish pumping engines but also for sewage-pumping mill engines in Lancashire (the last was operating until 1965), in marine steam engines and railway locomotives.

Woolf retired from Harvey's in 1833 and died in the Channel Islands in 1837. Working examples of Woolf's compound beam engines can still be seen at Abbey Pumping Station in Leicester, Claymills Pumping Station in Burton-upon-Trent and Blagdon Lake in North Somerset.

A Snorting Monster
Goldsworthy Gurney's steam road carriage

Born in Padstow in 1793, Cornishman Goldsworthy Gurney became one of those larger-than-life geniuses who came to symbolise the Victorian age of entrepreneurship and scientific advancement. He became a doctor at the age of 20 years before moving to London in 1820. Here, he gave lectures on chemical science and also applied his practical skills to building a steam carriage and developing the oxyhydrogen blowpipe and the steam blastpipe.

Gurney's steam carriages were built at Regent's Park in London between 1825 and 1829. Looking very similar to the traditional stagecoach, they were steered by a small pair of wheels at the front with

Gurney's steam road carriage

the enclosed steam engine in the rear. His carriages, capable of 20mph, were often seen around North London and, in 1829, he drove one to Bath and back along the route of what is now the A4 at an average speed of 14mph. Although this was 11 years before the Great Western Railway reached the city, the public were not impressed – riding on a steaming, snorting monster like this was not their idea of safe travel! To allay their fears Gurney built an articulated steam carriage with the steam engine (known as a steam drag) separated from the passenger compartment. After being transported there by sea, one of these ended up in Glasgow and its remains were given to the Glasgow Museum of Transport (now the Riverside Museum where it can be seen today) in 1889.

An Illiterate Genius
George Stephenson – the father of all railways

George Stephenson was born in Wylam, Northumberland, in 1781. The second son of illiterate parents, George went to work as an engineman at a colliery in Newburn in 1798 while at night he went to school to learn the three Rs. In 1801 he became a brakesman controlling pit winding gear, and he married the following year. George's son Robert (see 'Like Father, Like Son') was born in 1803 but, sadly, his wife Fanny died in 1806 from TB. By 1811, George had become well known for his skills in maintaining and repairing stationary steam colliery winding and pumping engines and was appointed engine wright at Killingworth Colliery.

Richard Trevithick (see 'The First Road Locomotive') had already built a steam locomotive for the Wylam Colliery near Newcastle in 1805 but it had proved too heavy for the wooden waggonway on which it ran. Spurred on by Trevithick's work, George designed and built his first steam locomotive, named *Blücher*, to haul coal wagons on the Killingworth Colliery waggonway. Capable of hauling 30 tons of coal uphill at 4mph it was the first successful flanged-wheel steam adhesion locomotive in the world and also the first to have cylinder rods that connected directly to the wheels. Several similar locomotives were built for other collieries and

George Stephenson's Blücher

the Kilmarnock & Troon Railway in Scotland. However, all of them suffered from one problem – they were all too heavy for the flimsy track that they ran on.

George went on to design improved cast iron rails that were used with some success on the Hetton Colliery Railway near Sunderland that he built in 1820. Again this was a world first as it was the first railway not to use horses, instead employing only gravity and steam haulage.

Following his success at Hetton Colliery, George (along with his 18-year-old son, Robert) was appointed to survey a new route for the planned Stockton & Darlington

Railway and also to supply steam locomotives for the new line. In partnership with the SDR's Director Edward Pease, George established a locomotive factory in Newcastle. Known as Robert Stephenson & Co., its Managing Director was George's son, Robert.

The company built four locomotives for the SDR and the first, *Locomotion No. 1*, was driven by George on the SDR's grand opening day on 27 September 1825. This was to be a day of world firsts, as the SDR was the world's first public railway, and *Locomotion* was the first steam locomotive to use coupling rods connecting the driving wheels.

Stephenson had also set the gauge of the railway to 4ft 8in and with a half-inch added on later this has since become the industry standard across most of the world. *Locomotion* was a great success, hauling an 80-ton train of coal wagons and a passenger coach full of dignitaries (again the first in the world) over the nine-mile line in two hours.

George Stephenson's success with the Stockton & Darlington Railway soon led to him being appointed engineer for the planned Liverpool & Manchester Railway. In conjunction with his son, Robert, he designed a steam locomotive, *Rocket*, to take part in the Rainhill Trials held in 1829.

The Rainhill Trials
A railway Grand Prix

The Rainhill Trials were held to find the most suitable form of motive power for the new Liverpool & Manchester Railway. Five locomotives took part including Robert Stephenson's *Rocket*, Ericsson and Braithwaite's *Novelty* and Timothy Hackworth's *Sans Pareil* (see 'The First Locomotive

Superintendent'). *Rocket*, built at Robert Stephensons factory in Newcastle, was the only locomotive to complete the trials and won £500 for its designers and builders. This groundbreaking locomotive had several new innovations – an efficient multi-tube boiler, a blastpipe and two cylinders set at an angle driving a pair of driving wheels. At the trials the locomotive hauled a 13-ton load at an average speed of 12mph with a maximum speed of 29mph. Robert Stephenson & Co. was also awarded the contract to build locomotives for the LMR.

The Liverpool & Manchester Railway opened in grand style on 15 September 1830. The opening ceremony of the world's first inter-city railway was attended by the great and the good including Prime Minister Arthur Wellesley, Duke of Wellington. The leading locomotive *Northumbrian* was driven by George Stephenson – a vast improvement on the *Rocket*, it featured a firebox fitted to the rear of the boiler and a smokebox door on the front. By 1830 steam locomotive development was proceeding at such a pace that

George Stephenson's Rocket

within less than a year Robert Stephenson's *Planet* (see 'Like Father, Like Son') took another major technological leap forward featuring, for the very first time, inside cylinders and a steam dome.

By now George had become a wealthy man and his skills as a railway and locomotive engineer were soon in demand from fledgling railway companies at home and abroad. In recognition of his enormous contribution to railways he was elected the first President of the Institute of Mechanical Engineers in 1847, a Society that today boasts around 80,000 members worldwide. Soon after marrying for the third time George died at his home, Tapton House, Chesterfield, in 1848. He is buried in Holy Trinity Church, Chesterfield.

STEAM FIRE ENGINES

More famous for his partnership with John Ericsson in building the steam railway locomotive Novelty *for the 1829 Rainhill Trials (see 'The Rainhill Trials'), John Braithwaite also built the world's first practical steam fire engine in 1830. Although horse-drawn, the rear-mounted steam engine with a vertical boiler fuelled by coke drove a powerful water pump and was successfully used to put out serious fires at the English Opera House and the Houses of Parliament in the 1830s. By the mid-1860s three British manufacturers had cornered the market for these horse-drawn fire engines – Shand Mason of Blackfriars, Merryweather & Sons of Greenwich and William Rose of Manchester. Capable of raising steam from cold water in less than ten minutes they were sold in their hundreds to fire brigades throughout Britain and also exported to Europe and the colonies. Both Merryweather and Shand Mason introduced self-propelled steam fire engines in the early 20th century but the introduction of motorised fire engines by World War I soon led to their demise. Restored examples can regularly be seen in action at Steam Fairs around the UK and several are also exhibited at the London Fire Brigade Museum in Southwark.*

The First Locomotive Superintendent
Timothy Hackworth

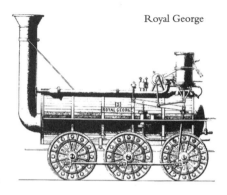

Royal George

Timothy Hackworth was born in 1786 in Wylam, Northumberland, George Stephenson's birthplace just five years earlier. Following in his father's footsteps, Hackworth became a boilermaker at Wylam Colliery in 1807. Working closely with William Hedley (see 'From Waggonway to Railway'), he played a major part in the design, construction and the continuing maintenance of *Puffing Billy* which was built in 1813.

In 1825 Hackworth was appointed Locomotive Superintendent of the new Stockton & Darlington Railway and while in this post he made important modifications to Stephenson's locomotives by realigning the steam blastpipe,

thus bringing about a marked improvement in locomotive performance. Built by Robert Stephenson & Co. in 1827 for the SDR, the 0-6-0 *Royal George* incorporated many of Hackworth's ideas and subsequent locomotives built at Newcastle all featured his innovative blastpipes.

Hackworth also entered his own locomotive, the *Sans Pareil*, for the Rainhill Trials on the Liverpool & Manchester Railway in 1829. Although not the winner his locomotive, featuring vertical cylinders and a blastpipe, performed well before breaking down with a broken cylinder.

EARLY STEAM EXPORTS

While Locomotive Superintendent of the SDR, Hackworth also went into business with his son, John, building steam locomotives at Shildon, the western terminus of the railway. Their output included the first steam locomotive to run in Russia in 1836, and one of the first (0-6-0 Samson) to run in Canada in 1838. Hackworth continued to work on the development of steam locomotives until shortly before his death in 1850. His former home at Shildon is now a museum.

Like Father, Like Son
Robert Stephenson

Born near Wallsend on Tyneside in 1803, Robert Stephenson was the only son of railway pioneer George Stephenson and received the education that his father had missed out on. He began his apprenticeship at Killingworth Colliery, before joining his father and railway pioneer Edward Pease in 1823 when they founded the world's first railway locomotive works, Robert Stephenson & Co., in Newcastle.

However, Robert departed for South America the following year where he worked as a mining engineer in Colombia until 1827.

Returning to Newcastle, he took over the running of the company and played a major part in the design of the *Rocket* which won the Rainhill Trials on the Liverpool & Manchester Railway in 1829. The following year he designed the *Planet*, a groundbreaking 2-2-0 locomotive that was the first in the world to have inside cylinders and a steam dome.

In 1833 Robert's career took a different direction when he was appointed Chief Engineer of the newly authorised London & Birmingham Railway, the first main line to serve London. From that point on he excelled as a railway engineer and, in particular, as a bridge designer, not only in Britain but also in Europe, North Africa and North America. In Britain his most famous examples are the High Level Bridge in Newcastle, the Royal Border Bridge at Berwick and the former tubular Britannia Bridge in North Wales. Further afield his skills as a railway and bridge builder also took him to France,

Switzerland, Spain, Egypt and Canada.

In later years he became an MP and on his death in 1859 he was buried with full honours in Westminster Abbey.

Early Steam Locomotive Survivors

Where to see early steam locomotives

PUFFING BILLY

The Science Museum, Exhibition Road, South Kensington, London SW7 2DD
Tel: 020 7942 4000
Website: www.sciencemuseum. org.uk

Designed by William Hedley and Timothy Hackworth in 1813, *Puffing Billy* was the world's first adhesion steam locomotive and was given to the Patent Office Museum (later to become part of the Science Museum collection) in London in 1862. It is the world's oldest surviving steam locomotive and can be seen today in this location.

WYLAM DILLY

National Museum of Scotland, Chambers Street, **Edinburgh EH1 1JF**
Tel: 0300 123 6789
Website: www.nms.ac.uk

A sister engine of *Puffing Billy* (see above), *Wylam Dilly* was built in 1815 and is the world's second oldest surviving steam railway locomotive.

LOCOMOTION NO. 1

Head of Steam, Darlington Railway Museum, North Road Station, Darlington DL3 6ST
Tel: 01325 460532
Website: www.darlington.gov. uk/Leisure/headofsteam

Built by Robert Stephenson & Co. for the opening of the Stockton & Darlington Railway (the world's first public railway) in 1825, *Locomotion No. 1* was retired in 1841 when it became a stationary steam engine. In 1857 it was preserved and put on display. For many years it was a familiar sight at Darlington Bank Top station but in 1975 was moved to the Darlington Railway Centre and Museum, now Head of Steam.

ROCKET

The Science Museum, Exhibition Road, South Kensington, London SW7 2DD
Tel: 020 7942 4000

Website: www.sciencemuseum. org.uk

Much modified, George Stephenson's winning locomotive at the 1829 Rainhill Trials on the Liverpool & Manchester Railway ended its life working on Lord Carlisle's Colliery lines in Cumberland until 1840. It was given to the Patent Office Museum in London (now part of the Science Museum's collection) in 1862. At least four replicas were later built – one for the Henry Ford Museum in Michigan, USA, one for the Museum of Science & Industry in Chicago, USA, and two for the National Railway Museum in York. Of the latter, one is sectioned for display purposes while the other is operational.

SANS PAREIL

Shildon Locomotion Museum, Shildon, Co. Durham DL4 1PQ Tel: 01388 777999

Website: www.nrm.org.uk

Designed by Timothy Hackworth, *Sans Pareil* took part in the 1829 Rainhill Trials on the Liverpool & Manchester Railway. Although unsuccessful in the competition the loco went on to work on the LMR before ending its life on the Bolton & Leigh Railway. Along with a replica, it is now preserved at the Shildon Locomotion Museum, an offshoot of the National Railway Museum.

PLANET (replica)

Museum of Science & Industry, Liverpool Road, Castlefield, Manchester M3 4FP Tel: 0161 832 2244

Website: www.mosi.org.uk

Built by Robert Stephenson in 1830, the *Planet* was the first steam locomotive in the world to feature inside cylinders and a steam dome. A replica can be seen in operation at the Museum of Science & Industry in Manchester.

SAMSON

Nova Scotia Museum of Industry, 147 North Foord Street, Stellarton, Nova Scotia, Canada

Website: www.museum.gov. ns.ca/moi

Built by Timothy Hackworth for the Albion Mines Railway in Nova Scotia, Canada, *Samson* is the oldest surviving steam railway locomotive in Canada. It is now on static display at the Nova Scotia Museum of Industry.

LION

Museum of Liverpool, Pier Head, Liverpool, L3 1DG
Tel: 0151 478 4545
Website: www.liverpoolmuseums. org.uk

Built in 1838 for the Liverpool & Manchester Railway, *Lion* remained in service until 1858 when it became a stationary boiler in Mersey Docks. It remained there until 1928 when it was preserved and took part in the LMS Centenary Celebrations in 1930. The loco also played a starring role in the 1953 film *The Titfield Thunderbolt*, starring Stanley Holloway, Naunton Wayne, John Gregson and Sid James.

RICHARD GARRETT & SONS

Founded in 1778 in Leiston, Suffolk, Richard Garrett & Sons were initially agricultural engineers and went on to build portable steam engines, traction engines, ploughing engines, steam rollers, steam tractors and steam wagons. As with Aveling & Porter (see 'King of the Steam Rollers'), Richard Garrett & Sons also joined the Agricultural & General Engineers in 1919 – production ceased in 1932.

The First Steam Bus
Walter Hancock's Enterprise

Born in 1799 in Marlborough, Wiltshire, inventor Walter Hancock became an enthusiastic advocate of steam-powered road vehicles. By 1824 he had set up a factory at Stratford in East London where he built a number of vehicles including a steam car and the first steam omnibuses. Built in 1829 his first bus, the *Infant*, could carry ten passengers and by 1831 was in regular service between East London and the City. This bus went on to inaugurate the first London to Brighton service in 1832 – nine years before the railway opened – but came to grief when the driver deliberately jammed the safety valve in order to raise more steam and the boiler exploded.

Hancock's next steam bus was more advanced but took a crew of three to operate it – a driver, a boilerman and a stoker. Named *Enterprise*, it incorporated many innovations such as a steering wheel, suspension and chain drive and, in 1833, went into service between the City of London and Paddington.

The *Automaton*
London's first successful bus service

Hancock's third bus was the *Automaton*. Seating 22 passengers it could travel at speeds of up to 20mph and went into regular service between the City, Islington, Stratford and Paddington in 1836. According to Hancock's own figures it carried over 12,500 passengers during this time.

Despite the success of Hancock's steam buses, the operators of horse-drawn buses forced him off the road by successfully lobbying for swingeing road tolls for steam-driven vehicles. Hancock's opponents also made life difficult for him by covering his route with loose, broken stones. By 1840 he had given up, instead concentrating on developing lightweight steam engines for use on railways. He died in 1852. A working replica of his bus *Enterprise* can be seen at the Beamish Museum in County Durham.

CHARLES BURRELL & SON
Charles Burrell was born in 1817 in Thetford, Norfolk, and took over his family's agricultural implement company in 1836. He built his first portable steam engine in 1846 and by the end of the century the company was producing large numbers of agricultural steam traction engines along with steam road locomotives, road rollers, ploughing engines and steam wagons. Charles Burrell died in 1906 and the company went on to be run by his son Charles Jr until closure in 1928.

Swindon's Founding Father
Daniel Gooch

The youngest of three brothers, Daniel Gooch was born in Northumberland in 1816 and trained as an engineer at Robert Stephenson's Vulcan Foundry in Warrington and on the new London & Birmingham Railway. In 1837, (at the tender age of 21 years) he was appointed Superintendent of locomotive engines for Brunel's fledgling broad gauge Great Western Railway.

His first locomotive design was the 'Firefly' Class 2-2-2 which were introduced in 1840 – these were

Gooch's development of the earlier 'Star' Class 2-2-2 that had been built for the GWR by Robert Stephenson. At this time the GWR didn't have its own locomotive works so Gooch's first engines were actually constructed by a whole range of locomotive builders such as Nasmyth, Gaskell & Co. and Fenton, Murray & Jackson.

Gooch's next step was to set up the GWR's own locomotive works and in 1840 he had decided on a green field site near the village of Swindon in Wiltshire. Gooch's first broad gauge locomotive to be built at Swindon, the 'Iron Duke' Class 4-2-2 *Great Western*, emerged from the works in 1846. With their eight-foot diameter driving wheels and a speed of 80mph, these magnificent machines (albeit with modifications) remained in service until the very end of the broad gauge in 1892.

Gooch remained in his post as Locomotive Superintendent until 1864 when he resigned. The following year he was Chief Engineer during the laying of the first transatlantic telegraph cable from Brunel's SS *Great Eastern* (see 'A Transatlantic Vision'), and was appointed Chairman of the GWR at a time when the company was in financial difficulties. While on board the SS *Great Eastern* out in the Atlantic he was also elected as Conservative MP for Cricklade, a seat he held for 20 years, and in 1866 he was created a Baronet for the laying of the transatlantic telegraph cable.

Gooch's strong chairmanship of the GWR continued until his death in 1889 – he was even the first person to crawl through the gap between the Welsh and English sides of the Severn Tunnel during its construction in 1884.

Gooch's 'Iron Duke'

Broad Gauge Reborn
Reliving Brunel's broad gauge

Sadly, none of Daniel Gooch's broad gauge locomotives survived for long after the end of broad gauge on the GWR in 1892. One member of his 'Iron Duke' Class 2-2-2, namely *Lord of the Isles*, was initially preserved at Swindon Works along with *North Star* (see below), but was scrapped due to lack of space in 1906. Now, only its eight-foot diameter driving wheels have survived and can be seen at Swindon Steam Museum.

A non-working replica of Robert Stephenson's 'Star' Class 2-2-2 *North Star* was built in 1923 using some parts of the original locomotive which was also scrapped at Swindon in 1906. It can now also be seen at Swindon Steam Museum.

The Great Western Society has recreated a section of broad gauge railway at Didcot Railway Centre using materials recovered from a disused railway in Devon. They have also built a working replica of Daniel Gooch's 'Firefly' 2-2-2 locomotive that can be seen hauling a replica open carriage on certain running days during the year.

CLAYTON & SHUTTLEWORTH
Founded in 1842 in Lincoln, the partnership of Nathaniel Clayton and Joseph Shuttleworth started building portable steam engines for agricultural use in 1845. Used to power machinery around the farm such as threshing machines, the firm became one of the largest producers and exporters of portable steam engines in the world, having built around 30,000 by the end of the 19th century. In the early 20th century they started to produce agricultural machines powered by the internal combustion engine, but went bankrupt in the 1930s.

Steam in the Sky
The aerial steam carriage

Designed by William Henson and John Stringfellow in 1842, the Aerial Steam Carriage was one of the earliest experiments into powered flight. Both men were inventors and engineers working in the lace-making industry in Chard, Somerset and their patented design nearly never got off

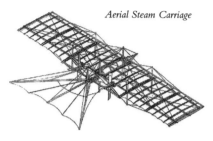

Aerial Steam Carriage

the drawing board. A small steam-powered model did, however, manage a short flight in a disused lace mill in Chard, supported on a length of wire. This is the first recorded occasion of powered flight in the world.

As patented, Henson and Stringfellow's full-size 'aircraft' had a total wing area of 6,000ft and was designed to be powered by a lightweight 50hp steam engine, carrying around 10 passengers at speeds of up to 50mph over a distance of 1,000 miles. Their company, the Aerial Transit Company, failed to gain any financial support for the project and even became the butt of jokes and cartoons. An exhibition of Henson and Stringfellow's work can now be seen in the Chard Museum, Somerset.

A Transatlantic Vision
Brunel's SS Great Western

One of the greatest British engineers of all time, Isambard Kingdom Brunel, was born in Portsmouth in 1806. Educated by his French-born father, the engineer Marc Isambard Brunel, the young Isambard became a child prodigy and by the age of eight had learnt the technique of drawing buildings, basic engineering principles, geometry and also spoke French fluently.

Later he received a first class education in France where he also studied as a clock-maker, returning to Britain to work with his father on the building of the Thames Tunnel in 1822. In 1831 he won a competition to build the Clifton Suspension Bridge in Bristol but work was soon halted by the infamous Bristol Riots. This famous landmark bridge was not completed until 1864, five years after Brunel's death.

In 1833, Brunel was appointed as Chief Engineer of the newly-formed Great Western Railway between London Paddington and Bristol Temple Meads. Built to a broad gauge of 7ft 0¼in, this superbly engineered railway line was opened throughout in 1841 and soon became a byword for speed and comfort. Always an innovative engineer, Brunel didn't always get it right, as witnessed by his short-lived atmospheric railway in South Devon which saw a short working life of only seven months in 1848. Even his railway broad gauge had become obsolete by 1892.

However, undeterred by a few failures, Brunel was also a visionary and saw his Great Western Railway as the first link in a transatlantic route between London and New York. To this end he had already been instrumental in the formation of the Great Western Steamship Company in 1835, following this up with a theory that larger ships were more economical to operate than smaller ones. And so his wooden paddle-wheel steamship, the appropriately named SS *Great Western*, was born – when launched on the River Avon at Bristol in 1837 she was the first steamship specifically built to carry passengers across the Atlantic.

TRAVEL IN STYLE

The 235ft-long, 1,340-ton SS **Great Western** *was powered by two steam engines built by Henry Maudslay (plus four auxiliary sails). She could carry 138 passengers (in style and comfort) and 60 crew between Bristol and New York at a top speed of 8 knots. Her first voyage from Avonmouth to New York in 1838 took 15 days, 12 hours, a journey time that was reduced to 13 days, 12 hours by May 1839. Soon, increasing competition from rival transatlantic steamship companies led to the building of a much larger vessel, the SS* **Great Britain** *(see 'A Lucky Survivor').*

The SS **Great Western** *was rebuilt in 1840 with increased accommodation and made 45 return Atlantic crossings before being sold to the Royal Mail Steam Packet Company in 1847 to work the Southampton to West Indies mail service. She served as a troop ship during the Crimean War before being scrapped in 1856.*

Archimedes Screw
Francis Petit Smith's Steam Screw Propeller

The technology for moving water had first been discovered by the Greek scientist Archimedes in the 3rd century BC. Known as the Archimedes screw, the device was a screw-shaped blade that revolved inside a cylinder and is first thought to have been used for removing bilge water from sailing ships.

By the 1830s in Britain, steam-powered ships were still propelled by paddle wheels but in 1835 the inventor Francis Petit Smith became the first person in the world to take out a patent for a screw propeller. A farmer by trade, Smith had always had a fascination for boats and had experimented with a model boat

The first trial screw propeller

driven by a wooden screw on a reservoir at his farm. Spurred on by the success of this experiment Smith went on to build a steam-powered canal boat, fitted with a wooden propeller, which he demonstrated on the Paddington Canal in 1837. The propeller on the boat, originally built with two screw turns, broke and Smith discovered that the resulting single screw turn was in fact much more effective and increased speed.

SS *ARCHIMEDES*
While Smith was conducting his experiments with screw propellers, Swedish-born John Ericsson had also demonstrated a screw-propelled ship on the River Thames. The Admiralty were not impressed so Ericsson took his design to the United States where he became the designer of the US Navy's first screw-propelled warship, USS **Princeton.**

Back in Britain, Smith redesigned his canal boat with a single screw iron propeller. He made the bold decision to test it at sea, taking it down the Thames and along the Kent coast where, in stormy seas, it performed beyond expectation

– his journey was witnessed by officers of the Royal Navy who were suitably impressed. Encouraged by this, Smith successfully obtained financial backing to build a much larger sea-going vessel. Appropriately named SS **Archimedes,** *this wooden ship was built in 1838 by Henry Wimshurst in East London. Weighing 237 tons it was powered by two low-pressure steam engines that drove a propeller shaft connected to Smith's patented iron propeller – the latter was nearly 6ft in diameter and consisted of a single screw turn. The ship was also fitted with sails – uniquely, the propeller could be retracted into the vessel when under sail. The SS* **Archimedes** *made its maiden voyage from the Thames Estuary to Portsmouth in 1839 at an average speed of 10 knots. At Portsmouth the Navy top brass were so impressed by the ship's turn of speed that they conducted trials the following year at Dover. By now, fitted with an improved half screw turn, two-bladed propeller the SS* **Archimedes** *outperformed the Navy's fastest paddle-wheel boats, a feat that led the Navy to build its first screw-propelled warship,* **HMS Rattler** *(see 'A World First'), in 1843. Following its naval trials, the SS* **Archimedes** *went on to sail around Britain and further afield to Europe, exciting keen interest at every port. She was also lent to Isambard Kingdom Brunel (see 'A Transatlantic Vision') who was then building the world's largest ship, the SS* **Great Britain** *– the success of the* **Archimedes'** *screw propulsion led Brunel to adopt it for his giant ship. The* **Archimedes** *was also the forerunner of commercial screw propulsion ships which, by the 1850s, had largely replaced the paddle wheel for ocean-going ships. Smith actually lost money on his groundbreaking venture and he returned to farming. Recognition eventually came when he was rewarded by Parliament for his invention, was appointed Curator of*

SS Archimedes

the Patent Office Museum (later to become part of the Science Museum) in London and, in 1871, received a knighthood. He died at South Kensington in 1874 and is buried at St Leonard's Cemetery, Hythe, Kent.

A World First
The story of HMS Rattler

Following the successful maiden voyage of Henry Bell's *Comet* (see 'The First Clyde Steamer') in 1812 the Royal Navy lost no time in building its first steam-powered warship. Powered by a low-pressure Boulton & Watt steam engine driving twin paddle wheels, HMS *Congo* was in reality a three-masted sailing schooner designed for exploration of the River Congo in West Africa. She was launched in 1816 but in her sea trials the steam power proved unsatisfactory and the machinery was subsequently removed.

The Navy went on to build more steam-powered paddle-wheel ships but their use was confined to tasks such as harbour tugs and Channel packet boats. The trials held at Dover in 1840 with Francis Petit Smith's experimental screw-propelled ship SS *Archimedes* (see 'SS *Archimedes*') soon led to the building of the Navy's first screw-propelled wooden warship, the sloop named HMS *Rattler*. The first screw-propelled steam warship in the world, she was launched in April 1843 (the US Navy's steam screw-propelled USS *Princeton* was launched five months later) and in trials, including a tug-of-war with a paddle-wheeled warship, she proved conclusively that screw-propulsion was superior to the paddle wheel.

Fitted with two double-cylinder steam engines, built by Henry Maudslay of Lambeth, producing a total of 200hp, HMS *Rattler* was capable of 10 knots while under steam, although as with other early steam warships she also carried sails. From 1845 until her withdrawal in

HMS Rattler

1856 HMS *Rattler* saw active service in many conflicts around the world.

A Lucky Survivor
Brunel's SS Great Britain

As part of his vision for a transatlantic highway between London and New York, Brunel's SS *Great Britain* was the first iron-hulled screw steamship to cross the Atlantic. Much larger and heavier than his previous *Great Western*, she was powered by two twin-cylinder inclined direct-acting steam engines driving a single screw propeller and was also fitted with six sails.

Launched at Bristol in 1843, she could carry 252 passengers, a crew of 130, and 1,200 tons of cargo at a top speed of 11 knots, and was in service on the Bristol to New York run by 1845. A year later she came to grief on a sandbank off Northern Ireland – the cost of re-floating her ruined the Great Western Steamship Company. In 1852 *Great Britain* was acquired by Gibbs Bright & Co. who refitted her with an extra deck, two funnels and three masts. Now capable of carrying 750 passengers, she was

used as an emigrant ship to Australia until 1876, making 32 return voyages during that period.

By now showing their age, *Great Britain*'s steam engines were removed in 1881 and she became a sailing ship transporting Welsh coal to California via Cape Horn. On her third trip she came to grief at the Falkland Islands and was sold as a coal and wool storage hulk in Port Stanley. In 1937 she was beached and abandoned – this was surely the end for this historic ship.

But it wasn't. Thirty-three years later she was secured on a huge floating pontoon in the Falkland Islands and towed 7,000 miles back to her birthplace, the Great Western Dockyard in Bristol. Since arriving back in Bristol in 1970 she has been completely restored and now attracts around 160,000 visitors each year.

The First King of Derby
Matthew Kirtley

Matthew Kirtley was born in Tanfield, Co. Durham, in 1813 and started his railway career at the age of 13 years by becoming a fireman on the newly-opened Stockton & Darlington Railway. After stints as an engine driver on the Liverpool & Manchester Railway and the London & Birmingham Railway he was appointed locomotive foreman of the Birmingham & Derby Junction Railway in 1839, rising to Locomotive Superintendent in 1841. The B&DJR became part of the Midland Railway in 1844, and Kirtley became its first Locomotive Superintendent based at its headquarters in Derby.

During his 30 years at Derby, Kirtley expanded the works and by 1851 it had started to build locomotives, the first of which were the 2-2-2 'Jenny Lind' type. Over the succeeding years Kirtley designed three classes of 2-4-0 express passenger locos and two large classes (totalling 552 locos) of 0-6-0 goods locos – 11 of these outside-frame locos survived into the 1940s. Kirtley died in office in 1873.

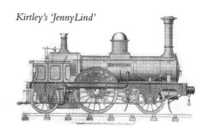

Kirtley's 'JennyLind'

and became an engine driver on the Liverpool & Manchester Railway. In 1843, he was appointed Locomotive Superintendent of the North Midland Railway, becoming an inspector of the newly formed Midland Railway a year later. Following a stint working on the construction of the Trent Valley Railway he was appointed Locomotive Superintendent of the London, Brighton & South Coast Railway in 1847, but died in the same year from a brain tumour.

The First King of Ashford
James Cudworth

THOMAS KIRTLEY
The elder brother of Matthew Kirtley (see 'The First King of Derby'), Thomas Kirtley was born in 1811

James Cudworth was born in Darlington in 1817; his two elder brothers became railway civil engineers. He began his own apprenticeship with Robert Stephenson & Co. in Gateshead in 1831, later gaining a permanent position as an engineer with the company. In 1840 he was appointed Locomotive Superintendent of the fledgling Great North of England Railway that was then building its main line between York and his hometown of Darlington.

Cudworth's big break came in 1845 when he was appointed Locomotive Superintendent of the breakaway South Eastern Railway. He oversaw the building of the company's new works on a green field site at Ashford in Kent which opened in 1847. He then went on to design numerous locomotive types including a class of 2-2-2 with 7ft driving wheels, 2-4-0 and 4-2-0 express locomotives, and was the first to introduce the 0-4-4 wheel arrangement for tank locomotives. Over 100 of his 2-4-0 coupled express locos were built between 1859 and 1875, of which two-thirds were constructed at Ashford. Also built at Ashford were over 50 of his standard goods 0-6-0 locos and in 1857 he patented a firebox that burnt

coal more efficiently without emitting unsociable clouds of smoke. Promoted to Locomotive Engineer of the SER in 1874, Cudworth soon fell out with the new Locomotive Superintendent, Alfred Watkin, over the secretive introduction of new locomotives designed by John Ramsbottom for the railway. He resigned in protest in 1876 and died in Reigate in 1899.

Brighton Moderniser
John Craven

J ohn Craven was born near Leeds in 1813. He first worked as an engineer on the Manchester & Leeds Railway and then the Eastern Counties Railway before being appointed Locomotive Engineer of the London, Brighton & South Coast Railway in 1847. He oversaw the modernisation and enlargement of the company's locomotive works at Brighton and designed a large number of locomotive types before retiring in 1869. In his latter years he acted as an adviser to canal and dock companies until his death in 1887.

The First King of Nine Elms
John Gooch

The second of three Gooch brothers, John was born in Northumberland in 1812. Following an apprenticeship under Joseph Locke during the building of the Grand Junction Railway, he was appointed resident engineer of that company on its opening in 1837. Three years later Gooch joined his elder brother Thomas, then resident engineer of the uncompleted Manchester & Leeds Railway.

In 1841 Gooch was appointed Locomotive Superintendent of the London & South Western Railway, a position he held until 1850. During his tenure at the LSWR he was not only instrumental in the opening of the company's locomotive works at Nine Elms in 1843 but also the design of several classes of single-wheeler express locos known for their turn of speed, and the 'Bison' Class 0-6-0s.

In 1850, Gooch was appointed Locomotive Superintendent of the Eastern Counties Railway, a position he held until 1856. During his time there he designed 2-2-2 well tanks, 0-6-0 tender locomotives and initiated locomotive construction at Stratford Works. Little is known about Gooch after he finally left the ECR and he died in 1900.

Rule Britannia!
The first screw battleships

Francis Petit Smith's SS *Archimedes* of 1839 (see 'Archimedes Screw') ushered in a new era of steam propulsion for ships. It wasn't long before more advanced types were being built for the Royal Navy.

HMS *Sans Pareil*

With the threat of war rumbling around northern Europe the Royal Navy launched its first steam-powered screw-propelled wooden battleship in 1851. Originally designed just to use sail power, HMS *Sans Pareil* was also fitted with a steam engine whilst under construction. In addition to carrying sails she was powered by a four-cylinder Boulton & Watt engine producing 500hp and saw active service around the world until her decommissioning in 1866.

HMS Agamemnon

HMS *Agamemnon*

When launched in 1852, the Royal Navy's HMS *Agamemnon* was the first British battleship to be designed and built from scratch with installed steam power. Also fully rigged with three masts, the 230-ft, 3,085-ton wooden ship was powered by a highly efficient 600hp steam trunk engine built by John Penn & Sons of Deptford, East London – this company had been the makers of oscillating marine steam engines for the Navy since 1844. So successful were Penn's high-pressure, high-revving engines that they were mass-produced for the Navy's fleet of gunboats used in the Crimean War.

HMS *Agamemnon* served in the Crimean War from 1853 to 1856 before helping to lay the first transatlantic telegraph cable

in 1858. She was decommissioned in 1862 and broken up in 1870.

The First King of Doncaster
Archibald Sturrock

Archibald Sturrock was born in Dundee in 1816, and became an apprentice at a local foundry in 1831. He later worked under Daniel Gooch (see 'Swindon's Founding Father') at the Swindon Works of the Great Western Railway where he eventually became Works Manager. In 1850, Sturrock was appointed as Locomotive Superintendent of the fledgling Great Northern Railway and oversaw the building of the company's new works at Doncaster that opened in 1853.

Here, Sturrock designed 28 different classes of locomotives, ranging from 36 of the 2-2-2 wheel arrangement and 52 of the 4-2-2 type, both used on express services between King's Cross and Doncaster, to 181 0-6-0Ts and 254 0-6-0 freight locos. The most long-lived of his designs were the LNER Class 'J3' 0-6-0s, some of which lasted until 1954, and the LNER Class 'J52' 0-6-0T which lasted until 1960. Having successfully overseen the vitally important first 15 years of the GNR, Sturrock retired in 1866 and died in 1909.

19th Century Survivors
Joseph Beattie and his well tanks

Joseph Beattie was born in Belfast in 1808 and trained as an architect before moving to England in 1835 where he became assistant to the engineer Joseph Locke during the building of the Grand Junction Railway and the London & Southampton Railway (later to become the London & South Western Railway). Following completion of the latter line in 1840, Beattie was appointed superintendent at the Nine Elms carriage and wagon works, becoming Chief Locomotive Engineer in 1850. During his tenure at Nine Elms, Beattie was responsible for the

development of the 2-4-0 locomotive for express work and the 2-4-0 well tank, introduced in 1874. Three of the latter locos (later classified as the '0298' Class) remained in service with British Railways until 1962.

THE SURVIVORS
Out of a total of 85 that were built, two of Beattie's vintage well tanks have been preserved: No. 30587 is normally based on the Bodmin & Wenford Railway in Cornwall; No. 30585 can usually be seen in action at the Buckinghamshire Railway Centre at Quainton Road. Joseph Beattie died suddenly in 1871 and was succeeded by his son, William, as Locomotive Engineer for the London & South Western Railway.

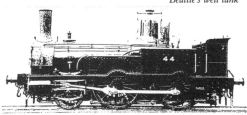

Beattie's well tank

Ploughing Furrows
John Fowler's steam plough

John Fowler was born in Melksham, Wiltshire, in 1826 and, forsaking his father's wishes, worked as an apprentice at the locomotive builders Gilkes, Wilson & Co. in Middlesbrough. Here, he gained experience building stationary steam engines and railway locomotives before building a successful steam-driven drainage plough in 1854. This was followed by his first steam-driven plough, built by Ransom & Sims in 1856, and a double-engine ploughing method in 1863.

A prolific inventor, by now Fowler had many patents to his name, and he went on to found John Fowler & Co. in Leeds the same year with his steam ploughing machines being sold worldwide. Sadly, John died in a riding accident at the age of 38 years in 1864, but his company lived on in the hands of his three brothers.

John Fowler & Co. continued to produce steam ploughing engines as well as traction engines, showman's engines, crane engines and steam rollers until the early 1930s. His beautifully decorated showman road engines were particular popular with operators of travelling fair grounds well into the 1950s.

LIMITED EDITION
Buckinghamshire agricultural engineer Thomas Rickett started to produce steam engines in 1857. A year later he built a steam-driven plough which so impressed the Duke of Sutherland that he ordered a three-wheeled steam carriage from Rickett in 1859. With a top speed of nearly 20mph, this chain-driven contraption could seat three passengers, one of them operating the steering, regulator and brake, and a boilerman at the rear. Costing the then princely sum of £200, a second model incorporating a gearbox was ordered by the Earl of Caithness

Steam plough

in 1860. It is not known whether any further models were made.

Crewe's Prolific Inventor
John Ramsbottom

'Lady of the Lake'

The son of a cotton mill owner, John Ramsbottom was born in Todmorden, Lancashire, in 1814. He first learnt about steam engines in his father's mill and at the local mechanics institute and, in 1839, went to work at the Atlas Works of locomotive builders Sharp, Roberts & Co. in Manchester. Here his boss, Charles Beyer (co-founder of Beyer-Peacock in 1854) was so impressed with his skills that he recommended the young Ramsbottom to the management of the newly opened Manchester & Birmingham Railway – Ramsbottom was appointed as Locomotive Superintendent of the M&BR in 1842.

The M&BR became part of the London & North Western Railway in 1846 and by 1857 Ramsbottom had risen to the position of Locomotive Superintendent at Crewe. Over the next 14 years a succession of successful locomotive types were turned out at Crewe, including nearly 1,000 of Ramsbottom's standard 'DX' 0-6-0 goods engines, 60 of the 2-2-2 'Lady of the Lake' Class and 260 of his 0-6-0 'Special Tank' Class – examples of the latter remained in service until 1959.

Also a prolific inventor, by 1880 Ramsbottom had accumulated 36 patents to his name including locomotive safety valves, locomotive hoists and water scoops. He retired from the LNWR in 1871 and died in 1897.

An Atlantic Monster
Brunel's SS Great Eastern

By comparison to his previous two ships *Great Western* and *Great Britain*, Brunel's final ocean-going vessel was a true monster for its age. The SS *Great Eastern* was conceived in 1851 as an emigrant ship on the North America run and was six times larger by volume than any other ship then afloat.

Built for the Eastern Steam Navigation Company between 1854 and 1858 at Millwall on the banks of the River Thames in east London, she certainly lived up to the description 'leviathan of the seas'. For the mid-19th century the statistics are staggering: weighing in at nearly 19,000 tons, this 692ft iron double-hulled monster was not only powered by a steam engine driving a 24ft-diameter four-bladed screw propeller and four other engines driving two 56ft-diameter paddles (each engine had its own funnel making five in all) but she also carried six enormous sails with a total area of 18,150sq ft. She was designed to carry 4,000 passengers and a crew of 418 at a maximum speed of 14 knots and had the range to carry this cargo to Australia (where it was then believed there was no coal!) and back without refuelling.

Built by Scott Russell & Co. she was launched sideways into the Thames on 31 January 1858 – three previous attempts had failed and the fourth only succeeded by using powerful state-of-the-art hydraulic rams. Before fitting out, the Eastern Steam Navigation Company became bankrupt and a new company, the Great Ship Company, was formed to operate the ship. After fitting out was completed, she started out on her maiden voyage around the south coast to Weymouth in September 1859. However disaster soon struck when a steam engine exploded, killing five stokers and throwing one of the funnels clear of the deck. The date was 9 September – six days later Brunel died from a stroke at the age of 53 years.

Following repairs, the *Great Eastern* finally made it across the Atlantic in 1860, taking 10 days 19 hours to

SS Great Eastern

complete the journey. Over the next three years she made a further nine return crossings of the Atlantic, but not without incident. In August 1861 she lost a paddle wheel and rudder in a storm and, a year later, she ran aground on rocks near New York.

A NEW LIFE

By 1864, the Great Ship Company (see 'An Atlantic Monster') was in serious debt and the **Great Eastern** *was put up for sale. She was bought for a fraction of her material value at an auction by a new company, the* **Great Eastern Steamship Company** *and was subsequently converted to lay undersea telegraph cables. Indeed, between 1865 and 1878 she laid over 30,000 miles of undersea cables across the Atlantic and Indian Oceans. After her illustrious career connecting far-flung corners of the British Empire by telegraph cable, the mighty* **Great Eastern** *met an inglorious end. The last ten years of her life were spent on the Mersey Estuary as a floating advertising hoarding for her new owners, a Liverpool department store. She was cut up at Rock Ferry on the shores of the Mersey between*

1889 and 1890 – not much now remains of this 'leviathan of the seas', but one of her masts was rescued for use by Liverpool Football Club and still proudly stands at their Anfield ground in the city.

Steam Luxury
The story of royal yachts

First introduced in 1823, private luxury steam yachts soon became popular with Victorian millionaires and royalty around Europe and the Mediterranean. All British royal yachts were owned and operated by the Royal Navy and early vessels were twin paddle steamers that also carried three sails.

Built for Queen Victoria, the first of these was the 1,000-ton twin paddle steamer HMY *Victoria and Albert*, built at Pembroke Dock and launched in 1843. She was replaced in 1855 by HMY *Victoria and Albert II* and renamed HMY *Osborne*, continuing in service to convey the British royal family to the Isle of Wight until being scrapped in 1868.

HMY *Victoria and Albert II* was

a much larger vessel launched in 1855, weighing 2,500 tons and with over 200 crew members. She was scrapped at the beginning of the 20th century, although a similar yacht was built for the Khedive of Egypt in 1865 and is still kept in working order by the Egyptian Navy.

The third HMY *Victoria and Albert* was much larger – she weighed in at 4,700 tons and was screw propeller driven. Although Queen Victoria died just before the ship was completed at Pembroke Dock in 1901, it served as royal yacht for Edward VII, George V, Edward VIII and George VI before being withdrawn from service in 1937. She survived World War II and was scrapped in 1954.

The final, and last, royal yacht was HMY *Britannia* and she was built at John Brown's shipyard on Clydebank. Weighing just over 5,700 tons and powered by a 12,000hp Parsons' steam turbine she was commissioned in 1954 and spent the following 43 years carrying the Queen and members of the royal family on official trips around the world, travelling over one million miles during her lifetime. She was

HMY Victoria and Albert

withdrawn from service in 1997 after carrying the Prince of Wales back to Britain at the end of British rule in Hong Kong. Since decommissioning she has become a tourist attraction at Leith on the Firth of Forth.

The First Ironclad
HMS Warrior

Until 1860 warships had been built of wood, but in that year the Royal Navy launched its first iron-hulled, armour plated, screw-driven steam warship, HMS *Warrior*. Also carrying three sails, this 418ft-long, 9,210-ton warship was fitted with a jet-condensing, horizontal trunk, single expansion steam engine built by John Penn & Sons. The ship was Britain's answer to the new French ironclad battleship, *La Gloire,* and signalled the start of an arms race between the two countries.

Capable of 14 knots, HMS *Warrior* was a major leap forward in warship design and, when under sail, her propeller could be retracted and the funnel lowered. Never seeing active service, her groundbreaking design soon became obsolete with the rapid advances in marine technology and she spent many years at Portsmouth as part of a naval training school. Unable to sell her for scrap, the Royal Navy towed her to Pembroke Dock in 1929 where she became a floating oil jetty for the next 50 years. Towed to Hartlepool for restoration in 1979, *Warrior* has been open to the public at Portsmouth since 1987.

Landlocked in Scotland
Scottish loch steamers

By the mid-19th century the coming of the railways to the Scottish Highlands and the English Lake District attracted Victorian tourists to these popular locations in their droves, all eager to enjoy the fabulous scenery of the mountains, lochs and lakes. At least five freshwater

HMS Warrior

lochs and three English lakes enjoyed a regular steamer service into the 20th century. As on the Clyde, the majority of these steamer services connected with railway-linked piers.

The first steamer on Loch Awe, the SS *Eva*, began service in 1861. She was followed into service on the loch by *The Lady of the Lake* which ran from 1863 to 1876, SS *Loch Awe* who ran from 1876 to 1914 and the TSS *Caledonia* who ran from 1895 until 1918. The latter ship was built at Paisley and was transported in sections by rail to the loch where she was assembled. The SS *Countess of Breadalbane* also saw service on the loch from 1882 to 1936.

In the Trossachs, Loch Katrine's popularity with tourists no doubt stemmed from its association with Rob Roy, Sir Walter Scott and Queen

Victoria. The first steam-powered boat on the loch, the paddle steamer *Gypsy*, began service in 1843 and was subsequently replaced by another paddle steamer in 1845, *Rob Roy*. However pride of place on the loch today goes to the SS *Sir Walter Scott* which was built by Denny's of Dumbarton in 1899. Her journey to land-locked Loch Katrine was a major feat – carried in sections on barges up Loch Lomond then transported overland by a team of horses she was reassembled at Stronachlachar. Powered by a three-cylinder triple expansion steam engine, the ship has two locomotive-type boilers that were fired by coke until 2007 when they were converted to run on bio-fuel. Today the *Sir Walter Scott* sails up and down this beautiful loch between April and October.

Much nearer the conurbation of Glasgow, Loch Lomond has always been very popular with day trippers. In 1818 the first loch steamer named the *Marion* arrived and since then a succession of steamers have plied up and down the loch. The last, the paddle steamer *Maid of the Loch*, was introduced by British Railways in

1953 and continued service until 1981. The very last paddle steamer to be built in Britain, she is powered by a two-cylinder compound diagonal steam engine and has a capacity for 1,000 passengers. She is now moored as a floating restaurant at Balloch awaiting full restoration for service on the loch once more.

On beautiful Loch Tay the paddle steamer *The Lady of the Lake* was introduced to connect with the branch line to Killin and Loch Tay station that opened in 1886. Day trippers from Glasgow were treated to a circular tour by rail, steamer, charabanc and rail, returning to Glasgow by rail from Aberfeldy. The steamer service was eventually withdrawn in 1949.

With the opening of the Caledonian Canal between Inverness

The Maid of the Loch

and Banavie (near Fort William) in 1822, Loch Ness became an important transport artery. Until the advent of modern roads the regular steamer service up and down the loch provided the only link with the outside world – one of the earliest vessels on the route was the paddle steamer *Glengarry* which was built in 1844 and remained in service along the loch and Caledonian Canal until 1927. Another paddle steamer, *The Gondolier,* was introduced in 1866 and spent her whole life working along the canal until the service ended in 1939.

Puffing Around Scotland
The story of Clyde Puffers

These delightful little Scottish steam workhorses were a regular sight transporting coal and goods around the west coast and Hebridean islands for nearly 100 years. First built for use on the Forth & Clyde Canal in 1856 they got their name from their distinctive puffing sound – drawing fresh water from the canal and powered

by a simple steam engine with no condenser, the exhausted steam puffed up the funnel as the piston stroked. Later models for seagoing use were fitted with a condenser but, although they had no puffing sound, the nickname stuck. Hundreds of these little boats were built although steam propulsion was replaced by diesel after World War II. Puffers built for Admiralty use during World War II were known as 'Victualing Inshore Craft', or 'VIC' for short.

Only a few coal-fired puffers have survived – *VIC 32*, based on the Crinan Canal, now gives cruises up and down the Caledonian Canal. The oldest surviving steam puffer, *VIC 27* (now named *Auld Reekie*), is currently being restored. One of the last working puffers, *VIC 72* (now renamed *Eilean Eisdeal*) is on display at the Inverary Maritime Museum.

Clyde Puffer

THE RED FLAG ACT

By 1860, steam road vehicles were seen by owners of horse-drawn vehicles as a threat to their livelihood. In 1865 Parliament gave into the horse lobby and passed the Locomotives in Highways Act. Otherwise known as the Red Flag Act it stipulated that all mechanically-powered road vehicles must:

* Have three drivers
* Not exceed 4mph on country roads and 2mph in towns
* Be preceded by a man on foot waving a red flag

(An amended Act of 1878 did away with the red flag nonsense but stipulated that steam road vehicles should 'consume their own smoke'.) The Red Flag Act, of course, virtually spelt the end for steam road vehicles in Britain until the Act was withdrawn in 1896 and the speed limit increased to 14mph. However, by this date the internal combustion engine was already being developed and steam power, particularly for smaller vehicles, never recovered.It was of course a different story in the USA where the famous Stanley steam car ruled the roads until the advent of the Ford Model T in 1908. One of Stanley's steam cars achieved a speed of 127mph in 1906 – a record for a steam car that was only finally exceeded in 2009.

King of the Steam Rollers
Aveling & Porter

Founded by Thomas Aveling and Richard Porter in Rochester, Kent in 1862, Aveling & Porter went on to produce steam wagons, traction engines and steam ploughing engines. By far their biggest success was their steam roller which was first produced in 1865 and sold in large numbers around Britain and abroad. In 1919, the company was one of the founder members of a combine made up of British engineering companies. Known as Agricultural & General Engineers, it failed to weather the onslaught of the internal combustion engine and went bankrupt in 1932.

Steam roller

Great Northern Racers
Patrick Stirling and his single-wheelers

Patrick Stirling was born in Kilmarncock in 1820 and served his apprenticeship at Blackness Foundry, manufacturers of mill and textile machinery, in Dundee before working for Neilson's Locomotive Works in Glasgow where he became foreman. In the late 1840s he became Superintendent of the Caledonian & Dumbartonshire Railway which opened in 1850. In 1853, Stirling was appointed Locomotive Superintendent of the Glasgow & South Western Railway which, the following year, moved its locomotive works from Glasgow to Kilmarnock. Here, Stirling designed 15 locomotive classes, the majority of which were of the 2-2-2 and 0-4-2 wheel arrangement, featuring his trademark domeless boilers.

In 1866, Stirling moved to Doncaster where he was appointed Locomotive Superintendent of the Great Northern Railway. During his 30 years in this post he designed many classic locomotive designs of which his 4-2-2 'Stirling Singles' are the most famous, giving outstanding performances on the GNR main line from King's Cross to Doncaster during the 'Railway Races to the North'. Fifty-two of these locos were built at Doncaster between 1870 and 1895 and No. 1 is now preserved at the National Railway Museum. A total of 28 different classes of locomotive were designed by Stirling before his death in office in 1895.

Lakeland Beauties
The story of English lake steamers

In the English Lake District the first public paddle steamer on Lake Windermere, *Lady of the Lake*, began service in 1845. Built of wood and powered by a 20hp steam engine,

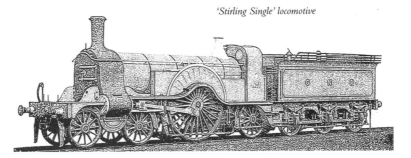

'Stirling Single' locomotive

she could carry 200 passengers and proved popular with tourists. The opening of the Kendal to Windermere railway in 1847 saw the introduction of additional steamers to the lake – the *Firefly* in 1847, followed by *Dragonfly* in 1850.

The local Furness Railway also got in on the act when in 1872 they took over the Windermere United Yacht Company, running the steamers to connect with their new railway terminus at Lakeside. Traffic continued to grow and in 1891 the railway company introduced a new steamer, the *Tern*. With a capacity for over 600 passengers, this steamer (albeit now diesel-powered) still plies up and down the lake. The last of the company's coal steamers, the *Swift*, began service in 1900. With a capacity for 781 passengers

she was the largest ship ever built for service on Windermere and was powered by steam compound engines. Refitted with diesel engines in 1956, she was retired in 1981 and finally broken up in 1999.

Also in the Lake District, on Ullswater the first paddle steamer, *Enterprise*, begin service in 1859. The second steamship, the *Lady of the Lake*, began service in 1877 after being transported overland in sections from Penrith. Converted to diesel power in 1935, during her long life she has suffered two sinkings and a fire but was restored and refitted before returning to service again in 1979 – she is now believed to be the oldest working steam passenger vessel in the world. Also still in service is the *Raven* which was launched on the lake in 1889

The Lady of the Lake

and was powered by a four-cylinder steam engine – she was also converted from steam to diesel in 1934.

On nearby Lake Coniston, the rebuilt steamboat *Gondola*, (originally built in 1859) is now operated by the National Trust.

Aston's Eccentric Inventor
The story of John Inshaw

Born in 1807, John Inshaw was an engineer and prolific inventor based in Aston, Birmingham. Among his many claims to fame are that he advised George Stephenson on the design of flanged wheels for his early steam locomotives, built a full-size working steam road carriage, and built the first twin-screw canal steamer. He was also present at the unsuccessful launch of Brunel's *Great*

Eastern (see 'An Atlantic Monster') at Millwall in 1858 and was the first to suggest the use of hydraulic jacks to move the giant ship into the Thames. Inshaw also ran a pub in Aston and filled it with working steam models including a steam-driven clock – he named the pub the 'Steam Clock Tavern'. He died in 1893.

Inshaw's twin-screw canal steamer, *Pioneer*, was powered by a small locomotive-type boiler that provided steam to two cylinders. These drove a crankshaft which, in turn, drove the propeller shafts via bevel gears. This vessel, the first of the classic narrowboats, proved highly successful in tests and led to more advanced types with a vertical flue boiler and a single-cylinder engine being used commercially on the Grand Junction Canal in the 1860s.

By 1876 the Grand Junction Canal owned 20 of these steam tugs (they also towed a barge or 'butty') that plied up and down the canal between London and Birmingham. However, following a catastrophic explosion caused by a stray spark igniting a load of gunpowder, these were all sold to the company that soon became known as Fellows, Morton & Clayton Ltd. Between 1912 and 1927 the company's steam tugs were gradually converted to be powered by simple two-stroke oil engines. However shorter steam tugs were still used by the Grand Junction Canal to tow boats through the Blisworth and Braunston Tunnels until the early 1930s. The last steam tugs were used on the Trent & Mersey Canal until 1943.

Inshaws also built a steam road carriage in 1881. This four-wheeled vehicle was powered by a high-pressure water tube boiler driving two cylinders that in turn drove the rear wheels via a double gear drive. It could carry ten passengers at a speed of around 10mph and was a regular sight around Aston until a law was brought into force prohibiting such vehicles.

Steam carriage

A New Broom at Brighton
The story of William Stroudley

William Stroudley was born near Oxford in 1833 and, at the age of 14 years, began an apprenticeship with steam engineer and inventor John Inshaw (see 'Aston's Eccentric Inventor') in Aston, Birmingham. In 1854, Stroudley left Inshaws to train as a locomotive engineer at the Great Western Railway's Swindon Works. This was followed by employment with the Great Northern Railway and the Edinburgh & Glasgow Railway before being appointed Locomotive Superintendent of the Highland Railway at Loch Gorm Works in Inverness. Here, he reorganised the works but only managed to design one class of diminutive steam locomotive – known as 'Lochgorm Tanks', they were the forerunner of Stroudley's famous 0-6-0T 'Terriers' which were later built for the London, Brighton & South Coast Railway.

Stroudley left the cash-strapped Highland Railway in 1870 when he was appointed Locomotive Superintendent of the London, Brighton & South Coast Railway at Brighton Works. Here, he introduced much-needed standardisation of locomotives and introduced six new classes including the 'D2' and 'B1' 0-4-2 express locos – the former saw service on the LBSCR's continental boat trains, while the latter, named after famous politicians, were on the London to Brighton run. However his 'A1' 0-6-0Ts (later rebuilt as 'A1X'), nicknamed 'Terriers', are the most famous with some remaining in service on the Hayling Island branch until 1963. Ten of these diminutive locos have been preserved to this day. Stroudley died in office in 1889.

Stroudley's 'Terrier'

A Somerset Steamer
Richard Grenville's steam carriage

Believed to be the oldest self-propelled passenger-carrying road vehicle still in working order, the Grenville steam carriage was built by Richard Neville-Grenville of Butleigh Court, near Glastonbury, around 1875. Grenville was assisted with the design by George Jackson Churchward who was to become the Chief Mechanical Engineer of the Great Western Railway in 1902.

Grenville's three-wheeled carriage was originally powered by a vertical boiler driving a single cylinder but this was later replaced by a two-cylinder arrangement. In addition to a driver it could carry four passengers with a boilerman at the rear and could reach a speed of 25mph. The carriage can be seen at the National Motor Museum, Beaulieu and also appears at vintage car events. In 2000, it successfully completed the London to Brighton Veteran Car Run in just over nine hours.

Grenville's steam carriage

A Double Life
Joseph Armstrong's mixed gauges

Joseph Armstrong was born in Cumberland in 1816 and grew up near Newcastle-upon-Tyne, within sight of the Wylam Waggonway, the stamping ground of William Hedley's *Puffing Billy* (see 'From Waggonway to Railway'). He first gained experience of steam engines whilst apprenticed at Walbottle Colliery and later worked as a driver on the Liverpool & Manchester Railway.

After following his brother to the Hull & Selby Railway in 1840, Joseph went on to become Assistant Locomotive Superintendent on the Shrewsbury & Chester Railway in 1847 – he was promoted to Locomotive Superintendent six years later. Following

amalgamation of the S&CR with the Great Western Railway in 1854, Joseph was appointed Locomotive Superintendent of that company's standard gauge fleet at Wolverhampton where, from 1859, GWR standard gauge locomotives were built.

In Swindon, Daniel Gooch (see 'Swindon's Founding Father') retired in 1864 and Joseph was appointed to replace him, with the new title of Locomotive, Carriage & Wagon Superintendent. During his tenure at the heart of 'God's Wonderful Railway', Joseph made his mark by being both paternalistic and strict with his workforce. A Methodist lay preacher, he also involved himself in the life of the fast expanding railway town of Swindon.

As a locomotive designer he led a double life, producing both broad gauge and standard gauge locomotives until his sudden death in 1877. Probably the most famous of his locomotives were the highly successful 'Metro' '455' Class 2-4-0 tanks that could be seen all over the system on suburban trains and the '310' '388' Class 0-6-0 goods engines, the precursor of the famous 'Dean Goods'. On his death he was succeeded by his brothers protégé at Wolverhampton, William Dean (see 'The Death of the Broad Gauge').

The King of Wolverhampton
The autonomy of George Armstrong

George Armstrong, younger brother of Joseph (see 'A Double Life') was born in 1822 and grew up near Newcastle-upon-Tyne. After an apprenticeship working with stationary steam colliery engines, in 1840 he and his brother were taken on as engineers for the newly opened Hull & Selby Railway where they worked as assistants to John Gray, the Locomotive Engineer. Gray was appointed Locomotive Superintendent of the London, Brighton & South Coast Railway in 1846 and George followed him down to Brighton. His next move was to France where he worked on the building of the Chemin de Fer du Nord before returning to Britain during the 1848 Revolution.

George then became a driver for the Shrewsbury & Chester Railway (where his brother was Assistant Locomotive Superintendent), rising to the rank of Locomotive Foreman. In 1854 the standard gauge S&CR was amalgamated with the broad gauge Great Western Railway and George, along with his brother, moved to Wolverhampton where the GWR had set up its standard gauge locomotive works. George became assistant to his brother, who had been appointed the company's standard gauge Locomotive Superintendent at the Stafford Road Works, and when Joseph was promoted to take over at Swindon in 1864, George stepped into his brother's shoes.

During his long reign at Stafford Road, George was virtually left to his own devices by his brother at Swindon and was responsible for the efficient building, rebuilding and repair of many types of saddle and side tank locomotives including 155 of the '517' Class 0-4-2 tanks, 130 of the '850' Class 0-6-0 saddle tanks, 60 of the '1016' Class 0-6-0 saddle tanks and 120 of the '1901' Class 0-6-0 saddle tanks. George had so much autonomy that even the Wolverhampton livery was noticeably different to the Swindon livery. He retired in 1896 after 33 years at the top and died, a single man, in 1901.

The Death of the Broad Gauge
William Dean's sojourn at Swindon

William Dean was born in South London in 1840 and, in 1855, became an apprentice under Joseph Armstrong (see 'A Double Life') at the Great Western Railway's Works in Wolverhampton. In 1863, Dean was appointed Chief Assistant to Armstrong and the following year, on the latter's promotion to Swindon, he was promoted to Manager of the Stafford Road Works under Joseph Armstrong's brother, George.

In 1868, Dean was further promoted as Chief Assistant to Joseph Armstrong at Swindon Works, taking over the coveted post of Chief Locomotive Superintendent of the GWR in 1877 after Joseph's untimely death. With a passion for the railway

town of Swindon and his beloved Wiltshire Regiment, he is best known for overseeing the end of Brunel's broad gauge in 1892, his experimentation with boiler designs, and the design of classic locomotive types such as the 'Duke' and 'Bulldog' Class 4-4-0s and his famous 'Dean Goods' 0-6-0 freight locomotives. Dean retired in 1902 but died in Folkestone only three years later.

ROBERT FAIRLIE'S PATENT
Scottish Locomotive Engineer Robert Fairlie (1830–1885) is best known for his double-bogie articulated steam locomotive that he patented in 1864. Built by George England of New Cross, London, Fairlie's novel locomotive was designed to operate on tightly curved, narrow gauge lines

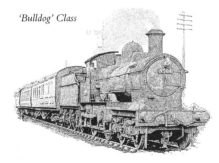

'Bulldog' Class

in mountainous regions and was first successfully demonstrated on the Ffestiniog Railway in 1870. Fitted with two boilers separated by a central driving cab, Fairlie's 0-4-0+0-4-0T **Little Wonder** *gave an outstanding performance on the steeply graded line before an invited audience of engineers from around the world. The demonstration was such a success that the Ffestiniog Railway went on to buy six Fairlie locomotives and also generated export sales. Today, preserved Fairlie locomotives can be seen in action on the Ffestiniog Railway which operates between Harbour Station, Porthmadog and Blaenau Ffestiniog in North Wales.*

A Highland Monster
David Jones and the first 4-6-0

David Jones was born in Manchester in 1834 and was apprenticed at the London & North Western Railway's Crewe Works. In 1855, he joined the Highland Railway at Loch Gorm Works, Inverness, working as an assistant to William Stroudley (see 'A New

Broom at Brighton') and Dugald Drummond (see 'Locomotive Longevity') before being appointed Locomotive Superintendent in 1870.

Jones designed five classes of 4-4-0 locomotives for the Highland Railway culminating in the 'Loch' Class which was introduced in 1896 – two examples remained in service for BR until 1950. However, he is most famous for his 4-6-0 'Jones Goods', the first loco class built to this wheel arrangement in Britain, which was introduced in 1894. Fortunately, the first of the class, No. 103, was saved for preservation in 1934 and now resides in the new Riverside Museum in Glasgow. Jones retired in 1896 and died in 1906.

The Giant of Crewe
The story of Francis William Webb

Francis Webb was born near Stafford in 1836 and started his apprenticeship with the London & North Western Railway at Crewe Works in 1851. Following this appointment, he worked his way up the LNWR ladder, becoming a draughtsman, Chief

Draughtsman, Works Manager and Chief Assistant to the Locomotive Superintendent, John Ramsbottom.

The latter retired in 1871 and Webb stepped into his shoes becoming Chief Mechanical Engineer, a position he held for over 30 years. During this period Webb designed 35 locomotive classes, many of them three- or four-cylinder compounds, which were built in huge numbers for Britain's 'Premier Line' at Crewe. The most numerous were his 500 0-6-0 'Coal Engines' introduced in 1873, 220 2-4-2T introduced in 1879, 310 0-6-0 'Cauliflower' introduced in 1880, 300 0-6-2 'Coal Tank' introduced in 1881, 500 'Special DX' 0-6-0 introduced in 1881, 166 2-4-0 'Renewed Precedent' introduced in 1887 and 170 0-8-0 four-cylinder compound heavy goods.

Also a prolific inventor with around 50 patents to his name, Webb resigned from the LNWR in 1903 and died in 1906.

An Illustrious Career
The story of Samuel Johnson

Samuel Johnson was born near Leeds in 1831 and began his apprenticeship with local locomotive builders E. B. Wilson & Co., makers of the famous 'Jenny Lind' 2-2-2 of 1847. In 1859, Johnson was appointed Acting Locomotive Superintendent of the Manchester, Sheffield & Lincolnshire Railway and five years later Locomotive Superintendent of the Edinburgh & Glasgow Railway. In 1866, he was appointed Locomotive Superintendent of the Great Eastern Railway at Stratford Works where he designed several 'firsts' in Britain including a 4-4-0 with inside cylinders and an inside frame, and a 0-4-4T.

Johnson's final move was to Derby where he was appointed Locomotive Superintendent of the Midland Railway in 1873, a post he held for 31 years. During his interregnum he was responsible for the introduction of many successful locomotive types of which the most famous are his beautiful 4-2-2 'Spinners' in 1896 (one is preserved

'Spinner' Class

at the National Railway Museum)
and the '1000' Class 4-4-0 three-
cylinder compounds introduced in
1902 (No. 1000 is also preserved).
However, his standard 0-6-0 goods
locos really stood the test of time
with a total of 935 being built, many
of which survived into the BR era
in the early 1960s. Strangely, for such
a numerically large class, none have
been preserved. After an illustrious
career, Johnson retired from Derby in
1904 and died in 1912.

GEORGE BRITTAIN

*Locomotive engineer George Brittain
was born in Chester in 1821. He
became Locomotive Superintendent
of the Caledonian Railway after
working in a similar position for the
Dundee, Perth & Aberdeen Junction
Railway from 1859 to 1863, and as
Assistant Superintendent for the*

*Scottish Central Railway from 1863
to 1865. He joined the Caledonian
as an assistant to Benjamin
Connor in 1865, taking over the
post of Locomotive Superintendent
in 1876. During his tenure at St
Rollox Works, Glasgow, Brittain
was responsible for the design and
construction of 85 outside cylinder
steam locomotives of various wheel
arrangements, ranging from 2-4-0
express locos to a diminutive 0-4-0
crane tank. He resigned in 1882 and
died soon afterwards.*

Lyme Regis Beauties
William Adams and his radial tanks

William Adams was born in
Limehouse, London, in
1823. Following apprenticeships in
the design of dockyards and steam
engines, he worked in Italy installing
steam marine engines in naval ships.
On his return to England in 1852,
Adams became a railway surveyor
before being appointed Locomotive
Engineer for the East & West India
Docks & Birmingham Junction
Railway (becoming the North
London Railway in 1858).

As Locomotive Engineer of the NLR he designed a series of 4-4-0 tank locomotives which were built at the company's workshops in Bow. In 1873 he became Locomotive Superintendent of the Great Eastern Railway at Stratford where he designed a series of tank engines, the 4-4-0 and 2-6-0 (the first in Britain) classes of locomotive. However, his main claim to fame at the GER was the modernising of Stratford Works.

Adams' third and final move as Locomotive Superintendent was to the London & South Western Railway's works at Nine Elms in 1878. During his 18 years there he designed eight classes of 4-4-0 express locomotives, 0-4-2 'Jubilees' as well as six classes of tank locomotives. Of the former his most famous and long-lived were the 4-4-2 radial tanks, the last three of which remained in service on the Lyme Regis branch until 1962. He also supervised the move of the carriage and wagon works from Nine Elms to Eastleigh in 1891. Adams retired in 1895 and died in Putney in 1904.

Radial tank

A LONE SURVIVOR
One example of Adams' radial tanks, No. 30583, has been preserved on the Bluebell Railway in Sussex where it is currently a static exhibit.

From Kilmarnock to Ashford
The story of James Stirling

James Stirling was born in Ayrshire in 1835 and went on to become an apprentice under his brother, Patrick Stirling (see 'Great Northern Racers'), who was Locomotive Superintendent of the Glasgow & South Western Railway at Kilmarnock Works. Following his apprenticeship, James worked for locomotive builders Sharp, Stewart in Manchester before returning to the GSWR where he eventually became Locomotive Superintendent following his brother's departure to join the Great

Northern Railway in 1866. During his stay at Kilmarnock, James Stirling was responsible for the design of 14 different locomotive classes, the majority of which were of the 2-4-0, 0-4-2 and 4-4-0 wheel arrangements.

In 1878, Stirling left Kilmarnock and moved south to Ashford, Kent, where he was appointed Locomotive Superintendent of the South Eastern Railway. During his 21 years in this post he was responsible for the design of six different classes of locomotive of which the most numerous were the 122 'O' Class 0-6-0, the 118 'Q' Class 0-4-4T and the 88 'F' Class 4-4-0. Stirling retired from the SER in 1898, the year before the company merged with the London, Chatham & Dover Railway, and died in 1917.

'Doon the Watter'
The golden age of Clyde steamers

Henry Bell's *Comet* of 1812 (see 'The First Clyde Steamer') was the first of hundreds of pleasure steamers that cruised down the Clyde from Glasgow and out to Largs, Campbeltown and Inverary and further afield until the 1960s. From

their very early days the locally-built Clyde steamers were aimed at the holidaymakers – the workers of industrial Glasgow who were eager to escape the satanic mills and to 'get-away-from-it-all', even if just for a day, 'doon the watter'.

Soon the railways were getting in on the act and by the late 19th century companies such as the North British Railway and the Caledonian Railway had their own fleets of steamers with their trains connecting with the ships at piers up and down the river. This was a situation that lasted well after the nationalisation of the railways in 1948. By the end of the 19th century there were over 300 (mainly paddle) steamers operating on the Firth of Clyde – it was a cut-throat business with companies vying for passengers by reducing fares and speeding up

PS Waverley

services. However the Clyde, by now an extremely busy stretch of water, was about to witness another world first in the history of steamships.

First demonstrated in Britain by Charles Parsons (see 'A Demonstration That Shook the World') in 1884, the steam turbine was a major step forward in the design of marine steam engines. Here, the tried and tested but complicated and inefficient reciprocating parts of compound steam engines were replaced by simpler and more efficient rotating parts which took their source of power from a high-pressure steam engine driving turbine blades. Built as a Clyde excursion steamer in 1901, the TS *King Edward* was the world's first commercial ship to be driven by steam turbines. With a maximum speed of around 20 knots, the ship was powered by a high-pressure steam turbine turning a single screw propeller shaft – the latter wasn't new, though, as screw propellers had first been introduced in 1839 (see 'SS *Archimedes*').

Launched by Charles Parsons' wife, the ship entered service by taking holidaymakers from railway-linked

TS King Edward

jetties along the Firth of Clyde and across the water to Campbeltown on the Mull of Kintyre. The TS *King Edward* was an immediate success and proved popular with holidaymakers, serving in both World Wars as a troop ship, and remaining in service until 1951. Although scrapped in 1952, this milestone ship's turbine engines were preserved and can be seen at the new Riverside Museum in Glasgow.

The most famous of the Clyde steamers is undoubtedly PS *Waverley* which has been preserved in working order and is the last seagoing passenger-carrying paddle steamer in the world. She was built in 1946 for the London & North

Eastern Railway, replacing an 1899-built ship of the same name that was sunk during the evacuation of Dunkirk in 1940. The current *Waverley*, powered by a triple expansion steam engine and capable of carrying over 900 passengers, operates passenger excursions during the summer months from ports around the coast of Britain.

A Cowlairs Man
Matthew Holmes' long-lived locomotives

Matthew Holmes was born in Paisley in 1844 and joined the Edinburgh & Glasgow Railway in 1859, eventually rising to the post of Locomotive Superintendent for the North British Railway in 1882. During his long tenure at Cowlairs Works, Holmes designed his famous Class 'C' 0-6-0 (LNER Class 'J36') – 188 were built, most of them at Cowlairs, with some examples remaining in service until the end of steam in Scotland in 1967. He also designed the Class 'M' 4-4-0 (LNER Class 'D31') with some examples surviving until 1952. Holmes retired

in June 1903 and died a few weeks later. As a mark of respect to a much-loved man his funeral train was hauled from Lenzie to Haymarket by one of his own NBR engines.

FOWLER'S GHOST
Built by Robert Stephenson & Co. in 1861, this unique broad gauge 2-4-0 locomotive was designed by Sir John Fowler for use on the new Metropolitan Railway in London. Built as a 'cut and cover' line under London's streets, most of the railway ran in tunnels where the locomotive was designed to operate without emitting smoke or steam. Its secret lay in its combustion chamber, located between the firebox and boiler, which was filled with an enormous number of heat-retaining fire bricks. These were designed to be heated by combustion gases, thus heating the boiler, while the locomotive was underground. However, in practice, there was no way of controlling the heat from the bricks (a dangerous situation) and after two unsuccessful trial runs the loco was dumped unceremoniously at Hammersmith, never to be seen again.

From Inchicore to Darlington
The short reign of Alexander McDonnell

Alexander McDonnell was born in Dublin in 1829 and gained experience of railways in England, Romania and Turkey before being appointed, in 1864, Locomotive Superintendent of the Great Southern & Western Railway at Inchicore Works in Dublin. Here, he set about modernising the works and introduced standardisation of locomotive parts.

His skills were recognised by the North Eastern Railway who appointed him Locomotive Superintendent in 1883. McDonnell's time in this post was short – he resigned the following year following altercations with management over his locomotive designs for the company. However, although his Class '38' 4-4-0s and Class '59' 0-6-0s were not well received by crews, McDonnell was able to introduce the standardisation that the NER badly needed. In later years he acted as a Railway Consultant Engineer until his death in 1904.

MARINE TRIPLE EXPANSION ENGINES

By the late 1880s the development of the triple expansion steam engine had transformed the design of naval ships and vastly increased the viability of ocean-going commercial shipping. Eradicating sail power once and for all, they brought a new era of reliability into steam engineering and, by the end of the century, were being used by the world's navies in their battleships and cruisers.

The size and power of the output of these engines enabled much larger ships to be built – Harland & Wolf built the largest in the world for the White Star Line. Massive 30ft-high engines producing 15,000hp and weighing 1,000 tons each were fitted into that company's new 45,000-ton transatlantic liners, of which **Titanic** *(see 'A Night to Remember') is the most famous. By comparison, one of the Royal Navy's largest battleships of the pre-'Dreadnought' period (see 'Britannia Rules the Waves') was* **HMS Hood** *– she was launched in 1891 with a displacement of 14,150 tons, 13,000hp engines and a top speed of 17.5 knots.*

Stratford's Gobblers
The short story of Thomas Worsdell

Thomas Worsdell was born in Liverpool in 1838. He was apprenticed at the London & North Western Railway's Crewe Works before joining the Pennsylvania Railroad in the USA as a Locomotive Engineer in 1865. Six years later Worsdell returned to Crewe where he worked under Francis Webb (see 'The Giant of Crewe') before being appointed Locomotive Superintendent of the Great Eastern Railway in 1882. During his sojourn at Stratford Works he introduced five classes of locomotive including 289 0-6-0 (later Class 'J15'), some of which remained in service until 1962, and 160 2-4-2T (later Class 'F4'), nicknamed 'Gobblers' for their high coal consumption, the last of which was withdrawn in 1956.

In 1885, Worsdell was appointed Locomotive Superintendent of the North Eastern Railway and over the next six years introduced 15 different classes of locomotive, many of them compounds that were later rebuilt as simple engines by his successors. He retired in 1890 and died in 1916.

Unglamorous But Functional
Matthew Stirling and the Hull & Barnsley

The son of Patrick Stirling (see 'Great Northern Racers'), Matthew Stirling was born in Kilmarnock in 1856. He served his apprenticeship with the Great Northern Railway in Doncaster and eventually became Chief Locomotive Superintendent of the Hull & Barnsley Railway in 1885, remaining in that post until 1922 when the company was taken over by the North Eastern Railway.

During his 38 years with the H&B, Stirling designed ten different classes of locomotive, the majority of which were freight types for hauling heavy coal trains, and all of which later saw service on the LNER with his Class 'N13' 0-6-2Ts surviving into BR ownership. His most powerful design was the Class 'Q10' 0-8-0 which was in huge demand during

World War I hauling coal trains from South Yorkshire to Hull Docks.

Matthew Stirling retired in 1922 and died in 1931.

A True Lancastrian
John Aspinall at Horwich

John Aspinall was born in Liverpool in 1851. At the age of 17 he started his apprenticeship with the London & North Western Railway at Crewe. In 1875 he became Works Manager at the Inchicore Works of the Great Southern & Western Railway in Ireland, becoming Locomotive Superintendent in 1882.

Aspinall was subsequently appointed Chief Mechanical Engineer of the Lancashire & Yorkshire Railway in 1886 where, at Horwich, he was responsible for the design and building of over 300 2-4-2 tank locos, 60 diminutive 0-4-0 saddle tanks, 230 0-6-0 saddle tanks (converted from 0-6-0 goods locos) and nearly 500 0-6-0 freight locos. All of these four classes were long lived with many surviving well into the BR era.

Aspinall went on to become General Manager of the L&YR in 1899, introducing Britain's first main line electric trains in the Liverpool area between 1904 and 1913. Also serving as president of the Institute of Civil Engineers and the Institute of Mechanical Engineers, Aspinall died in 1937 at the age of 85 years.

STEAM TRAMS
Horse-drawn trams were first introduced by the Oystermouth Railway in South Wales in 1807. However, they only became more widespread in Britain after an 1870 Act of Parliament was passed that allowed their tramlines to be built along roadways. Steam-powered trams were tried in a few cities – separate steam tram engines pulled double-deck passenger cars. Built by the Falcon Engine & Car Company (later to become Brush Electrical Engineering) of Loughborough, steam trams were introduced on the Moseley to Birmingham route in 1884. The trailer cars could seat 60 passengers while the separate tram engine could be driven from either end – they last ran in 1906 and were

Steam tram

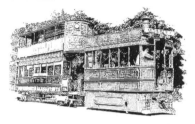

then replaced by electric trams. Steam trams were also used in Nottingham, West Bromwich, Accrington, Leeds, Huddersfield, Stockton and London but their activities were also soon cut short by the advent of the cleaner and more efficient electric trams.

In His Brother's Footsteps
Wilson Worsdell and the North Eastern Railway

The younger brother of Thomas Worsdell (see 'Stratford's Gobblers'), Wilson Worsdell was born in Crewe in 1850 and followed in his brother's footsteps as an apprentice at the London & North Western Railway's Crewe Works before moving to the USA where he worked as an engineer for the Pennsylvania Railroad until 1871. After a further period at Crewe Works, Wilson was appointed Assistant Locomotive Superintendent of the North Eastern Railway in 1883 before taking over from his brother as Locomotive Superintendent in 1890.

Wilson Worsdell's time at the NER was marked by his introduction of 27 different classes of locomotive including the powerful 'V' Class (later Class 'C6') 4-4-2 'Atlantics', and 'S' and 'S1' Class (later Class 'B13' and 'B14') 4-6-0s which gave such outstanding performances on Anglo-Scottish expresses between York and Newcastle. Another of his designs, the diminutive 'E1' Class (later Class 'J72') 0-6-0T introduced in 1898, was very long-lived with some examples surviving as station pilots at Newcastle Central until 1964. Uniquely, 28 of these 19th-century-designed locos were built by British Railways as late as 1949–1951. After a long and illustrious career at the NER, Wilson Worsdell retired in 1910 and died in 1920.

'J72' Class

Caley Flyers
John McIntosh and Cowlairs Works

A labourer's son, John McIntosh was born near Brechin, Scotland in 1846. By 1860 he had begun an apprenticeship with the Scottish North Eastern Railway at Arbroath before becoming a driver in 1867, by which time the railway had been absorbed by the Caledonian Railway. A 'Caley' man through and through, McIntosh worked his way up to become Locomotive Running Superintendent and, in 1895, was appointed Locomotive Superintendent at St Rollox Works in Glasgow.

McIntosh's reign at St Rollox was marked by a series of over 20 locomotive designs. These included the superb 4-4-0 'Dunalastairs' and the 4-6-0 'Cardeans' (both of which put in impressive performances on the CR main line between Carlisle and Glasgow) along with 'Oban Bogies' while his '2P' 0-4-4Ts and '2F' and '3F' 0-6-0 freight locos continued in service with BR into the 1960s. Honoured by King George V in 1911, he retired in 1914 and died in 1918.

20TH CENTURY STEAM BUSES
The repeal of the Red Flag Act in 1896 (see 'The Red Flag Act') reawakened interest in motorised road vehicles and, although the age of the internal combustion engine was about to dawn, experiments with steam vehicles restarted. Built by John Thornycroft, the first double-deck steam bus was introduced in London in 1899 and by 1909 the National Steam Car Company (the forerunner of the Eastern National Omnibus Company) was operating a small fleet of steam buses in London from their garage in Nunhead. By the beginning of World War I the company owned around 180 vehicles but these could never compete with the new

Steam bus

petrol-driven buses – the last steam bus ran in London in 1919. A later experiment by Sentinel (see 'Kings of the Road') was not successful – only four were built (with coachwork by Hora of Peckham) with three going to Czechoslovakia and the other used to carry the company's works brass band. A replica Sentinel steam bus based on an original Sentinel lorry chassis can be seen operating today from Windermere Steamboat Museum in the Lake District.

Great Eastern Flyers
James Holden at Stratford

James Holden was born in Whitstable, Kent, in 1837 and went on to serve an apprenticeship under his uncle, Edward Fletcher, who was then Locomotive Superintendent of the North Eastern Railway. In 1865, Holden moved to the Great Western Railway where he first became superintendent of the GWR workshops in Chester before becoming Chief Assistant to William Dean (see 'The Death of the Broad Gauge') at Swindon. After 20 years spent at the GWR

he was appointed Locomotive Superintendent of the Great Eastern Railway at its Stratford Works.

During his 23 years at Stratford, James Holden modernised the locomotive works, introduced the standardisation of locomotive parts, and designed and introduced over 20 different classes of locomotive. Probably the most famous of these are the 'Claud Hamilton' 4-4-0s, introduced in 1900, which were the mainstay of express motive power out of Liverpool Street until the 1930s, the Class 'T19' 2-4-0s, the 'S56' Class 0-6-0Ts (LNER Class 'J69') and the Class 'G58' (LNER Class 'J17') 0-6-0s which both remained in service for nearly 60 years, and the unique Class 'A55' 0-10-0 'Decapod'.

THE UNIQUE 'DECAPOD'

*Introduced in 1902, this 0-10-0
tank locomotive was built to prove
the rapid acceleration capabilities of
steam locos on suburban services in
response to a proposed new electric
railway. Although in tests it exceeded
all expectations (0 to 30mph in 30
seconds), the loco remained a one-
off and was converted to an 0-8-0
tender engine in 1906. The increasing
cost of coal also led Holden to
experiment with an oil-burning
locomotive as early as 1893. Known
as* **Petrolea,** *the converted Class
'T19' 2-4-0 was the first of over
100 oil burners using waste oil that
were subsequently built by the GER.
The paternalistic Holden held sway
at Stratford Works until 1907 when
his son, Stephen (see 'A Short, But
Spectacular, Reign') took over the
post. James Holden died in 1925.*

A Yorkshireman in Brighton
The story of Robert Billinton

Born in Wakefield, Yorkshire, in 1845, Robert Billinton served his apprenticeship at William Fairbairn's engineering company in Manchester. This was followed by employment between 1863 and 1870 at several large engineering firms and manufacturers of steam locomotives in Yorkshire.

In 1870 he moved to Brighton where he became Assistant to William Stroudley (see 'A New Broom at Brighton'), the newly appointed Locomotive Superintendent of the London, Brighton & South Coast Railway. In 1874 Robert moved again, this time to Derby where he became Assistant to Samuel Johnson, the Chief Mechanical Engineer of the Midland Railway.

William Stroudley died in 1889 and Billinton was appointed as his successor at Brighton. During his tenure as Locomotive, Carriage, Wagon & Marine Superintendent Robert Billinton designed many successful steam locomotive types (the 'B' Class 4-4-0, 'C2' Class 0-6-0, 'D' Class 0-4-4T, and numerous 'E' Class 0-6-2Ts) and steel-framed coaching stock for the LBSCR. He died suddenly in 1904 and was subsequently succeeded by Douglas Marsh (see 'Brighton Flyers').

Locomotive Longevity
Dugald Drummond's classic designs

Born in Ardrossan in 1840, Dugald Drummond went on to become one of the most prolific and successful steam locomotive designers of his time. Following an engineering apprenticeship in Glasgow, Drummond first worked as a boilerman for the civil engineer Thomas Brassey at Birkenhead before gaining employment at the Cowlairs Works of the Edinburgh & Glasgow Railway in 1864. He later moved to Inverness where he worked as a foreman under William Stroudley (see 'A New Broom at Brighton') at the Highland Railway's Loch Gorm Works. In 1870, Stroudley was appointed Locomotive Superintendent for

the London, Brighton & South Coast Railway at Brighton Works and Drummond went with him.

In 1875, Drummond moved back to Scotland to take the post of Locomotive Superintendent for the North British Railway and was involved as an expert witness during the official inquiry into the Tay Bridge disaster of 1879. While at Cowlairs Works Drummond designed seven classes of locomotive ranging from 0-6-0s and 2-2-2s to 4-4-0s and 0-4-2Ts, the majority of which remained in service through to the LNER era.

Drummond's next move was to the Caledonian Railway where he was Locomotive Superintendent from 1882 to 1890. Here he designed nine classes of locomotive, many of which survived into the BR era – his '294' Class 0-6-0 standard goods locos of 1883 were still being used in Scotland in the early 1960s, as were some of his '264' Class 0-4-0 saddle tanks.

From 1890 Drummond opened two locomotive engineering works, the first in Australia and the second in Glasgow, but returned to railway company employment in 1895

Greyhound 'T9' Class

when he was appointed Locomotive Engineer for the London & South Western Railway. Drummond oversaw the move of the LSWR's locomotive works from Nine Elms to Eastleigh in 1909. Until his death in 1912, he designed over 20 different classes of locomotives that ranged from the unusual 'E10' Class 4-2-2-0 to the famous 'T9' Class 4-4-0s or 'Greyhounds' as they became known. In addition to the latter class his '700' Class 0-6-0s and 'M7' Class 0-4-4Ts provided sterling service for both the Southern Railway and for British Railways until their demise in the early 1960s.

King of the Klondykes
Henry Ivatt's 'Atlantics'

Henry Alfred Ivatt was born near Ely in 1851 and started his apprenticeship under John Ramsbottom (see 'Crewe's Prolific Inventor') at the London & North Western Railway's Crewe Works in 1868. Ivatt went on to become a fireman and locomotive shed foreman with the LNWR before being appointed as Locomotive Engineer for the Great Southern & Western Railway at their Inchicore Works in Dublin in 1886. While in Ireland he patented the famous carriage window-opening device operated by a leather strap that stayed in service through to BR days in the 1960s.

In 1895, Ivatt was appointed Locomotive Superintendent of the Great Northern Railway at Doncaster Works. His 'Klondyke Atlantic' 4-4-2 locos of 1898 were the first of this wheel arrangement built in Britain and proved to be capable of high speeds with heavy loads along the East Coast Main Line between King's Cross and Doncaster – the first to be built, *Henry Oakley,* is now preserved at Bressingham Steam

'Klondyke Atlantic'

Museum at Diss, Norfolk. With later modifications by Nigel Gresley (see 'The World Beater') the last examples remained in service until 1945. The longest surviving Ivatt design for the GNR were the Class 'J52' 0-6-0 saddle tanks, introduced in 1897, several of which remained in service for BR until 1961.

Henry Ivatt retired in 1911 and died in 1923.

Could Try Harder
Henry Hoy at Horwich

Henry Hoy was born in London in 1855 and at the age of 17 years began an apprenticeship at the Crewe Works of the London & North Western Railway. Following six years employment as a draughtsman in the LNWR drawing office, Hoy was appointed as assistant in the locomotive department of the

Lancashire & Yorkshire Railway at Miles Platting, Manchester.

He became Works Manager at the new L&YR's works at Horwich in 1886, and was promoted to Chief Mechanical Engineer following John Aspinall's (see 'A True Lancastrian') promotion to General Manager in 1899. Hoy's period as CME of the L&YR was not marked by any high points in locomotive design and he left in 1904 to become General Manager of Beyer, Peacock, a post he held until his early death in 1910.

Brighton Flyers
Douglas Marsh's 'Atlantics'

Douglas Marsh was born in Aylsham, Norfolk, in 1862 and began his apprenticeship with the Great Western Railway at Swindon, rising to Assistant Works Manager by 1888. In 1896, he was appointed Chief Assistant Mechanical Engineer for the Great Northern Railway at Doncaster where he worked under Henry Ivatt (see 'King of the Klondykes') until 1904 when he left to join the London, Brighton & South Coast Railway as Locomotive

Superintendent at Brighton Works.

During his period at Brighton, Marsh introduced the Class 'H1' and 'H2' 4-4-2 express locomotives, no doubt with experience gained by working under Ivatt at Doncaster. The locomotives were soon employed hauling the crack LBSCR Pullman trains and were later rebuilt with superheaters by Richard Maunsell (see 'A Little Irish Magic'). The last member of this class was withdrawn in 1951. Marsh also designed a series of 4-4-2T and 4-6-2T along with a class of 0-6-0 goods engines and rebuilt a whole series of Stroudley and Billington locos. The most famous of these are the diminutive 'Terrier' Class 'A1X' 0-6-0s and they remained in service on the Hayling Island branch until 1963.

A BLACK CLOUD

Although Marsh officially retired on health grounds in 1911, there was much unsubstantiated speculation that he had been forced to leave the London, Brighton & South Coast Railway after pocketing the money intended for a new locomotive. He died in 1933.

A Short, But Spectacular, Reign
Stephen Holden and his 'B12' 4-6-0s

The third son of James Holden (see 'Great Eastern Flyers'), Stephen Holden was born at Saltney, Cheshire, in 1870 and joined the Great Eastern Railway as an apprentice at Stratford Works in 1886, one year after his father had been appointed Chief Mechanical Engineer. Here, Stephen worked his way up the GER ladder, first working as a draughtsman in the drawing office followed by promotion to inspector in the motive power department. In 1892 he was promoted to Suburban District Locomotive Superintendent, in 1894 to District Locomotive Superintendent at Ipswich, in 1897 to Divisional Locomotive Superintendent, then to Assistant Locomotive Superintendent before succeeding to his father as Locomotive Superintendent in 1908.

In this position he continued his father's work and introduced nine further classes of steam locomotives of which by far the most notable was the

'B12' Class

Class 'S69' 4–6–0 (LNER Class 'B12'). Introduced in 1911, over 80 of these fine locomotives were eventually built and gave sterling service during the LNER era not only in East Anglia but also in the Aberdeen area where they remained in service for BR until 1961. Fortunately, one member of the class has been preserved on the North Norfolk Railway.

Stephen Holden retired from the GER in 1912 and predeceased his father when he died in 1918 at the young age of 48 years.

Beauty of the Bens
Peter Drummond's classic design

The younger brother of Dugald Drummond (see 'Locomotive Longevity'), Peter Drummond was born in 1850 and became Locomotive Superintendent at the Highland Railway's Loch Gorm Works in 1896. During his 15 years here he designed four classes of locomotive of which the most well known are the 'Ben' Class 4-4-0s. Out of a total of 20 of these delightful engines, nine were built at Loch Gorm while the rest were built by Dübs & Co. and the North British Locomotive Company in Glasgow. Amazingly, ten survived to the BR era with the last, *Ben Alder*, being first saved for preservation but then scrapped.

Peter Drummond left the Highland Railway in 1911 to become Locomotive Superintendent of the Glasgow & Western Railway, a post he held until his death in 1918. The last example of his G&SWR 2-6-0 locomotives, known as Austrian Goods, just failed to survive into BR ownership as it was scrapped in 1947.

'Ben' Class

THE CRAIGIEVAR EXPRESS

Local postman 'Postie Lawson' built his own three-wheeled steam road vehicle in 1895 to help him with his rural post round. This eccentric machine has been restored to working order and can be seen at the Grampian Transport Museum, Alford, Aberdeenshire.

Up the Derby Ladder
Richard Deeley and the Midland Railway

Richard Deeley was born in Chester in 1855 and started his apprenticeship at the Midland Railway's Derby Works in 1875. Under the watchful eye of the MR's Locomotive Superintendent, Samuel Johnson (see 'An Illustrious Career'), Deeley slowly worked his way up the promotion ladder at Derby: Head of Testing Department in 1890; Inspector of Boilers, Engines & Machines in 1893; Works Manager in 1902; and Chief Electrical Engineer and Assistant Locomotive Superintendent in 1903. Johnson finally retired at the beginning of 1904 and Deeley was then promoted to Locomotive Superintendent of the Midland Railway.

During his six years in this post Deeley patented an improved valve gear and further developed the compound steam engine as originally designed by the Scottish engineer, Walter Smith. The result was the highly successful class of three-cylinder Midland Compound 4-4-0 locomotives and the rebuilt Johnson-designed 4-4-0s. Sadly, Deeley hit the proverbial buffer stop when the Midland management refused to allow the building of larger engines and he resigned in 1909. He continued to work as an engineer and by the time of his death in 1944 had 15 patents to his name.

JOHN THORNYCROFT'S TORPEDO BOAT

Built in 1876 by John Thornycroft (see 'Kings of the Road'), HMS Lightning was the world's first naval vessel to be armed with torpedoes. Just over 87ft long, she was powered by a 460hp two-cylinder compound steam engine that powered the vessel to a maximum speed of 18.5 knots and was used in various

*experiments at Portsmouth until
scrapping in 1896.*

A Demonstration That Shook the World
Charles Parsons' steam turbines

The son of an earl, Charles Parsons was born in London in 1854. His invention of the steam turbine in the late 19th century totally revolutionised marine transport, naval warfare and the generation of cheap and readily-available electricity.

Following a first class education in Dublin and Cambridge, Parsons began his career in Newcastle as an apprentice for the heavy engineering company of W. G. Armstrong in Elswick before progressing to become Head of Electrical Engineering at Clarke, Chapman & Co. in Gateshead. Here, in 1884, he developed a steam turbine that drove an electricity generator, an invention that is still used around the world to generate the vast majority of our electricity.

As an invention, Parsons' steam turbine is simplicity in itself – high-pressure steam (raised by boiling

water using any combustible fuel, whether it be coal, oil, gas or nuclear) drives a multi-bladed circular steam turbine, thus generating a high-speed rotary motion. This rotary motion can then be applied to driving a ship's propeller shaft or an electric generator. Compared to the steam reciprocating engine with all its moving parts, the steam turbine has a much greater power-to-weight ratio and is much more thermally efficient.

Parsons went on to develop the first compound steam turbine in 1887 and in 1889 founded his own company in Newcastle – his first commercial electric turbogenerator was installed in Germany in 1899.

Parson's greatest display of his revolutionary steam turbine came at the Spithead Navy Review held to commemorate Queen Victoria's Diamond Jubilee in 1897 when his experimental ship *Turbinia* took the world by storm. Built of steel and with a length of just over 100ft, the ship was powered by a coal-fired 2,000hp three-stage compound steam turbine driving three shafts, each of which had three propellers – a total of nine propellers.

Parson's Turbinia

THE CHEEK OF IT!

Twenty-sixth of June 1897 saw Queen Victoria's vast home fleet lined up for inspection in Spithead, the stretch of the Solent lying between Portsmouth and the Isle of Wight. In full view of the Prince of Wales (later King Edward VII), all the lords of the Admiralty and a throng of important foreign visitors, Parson's **Turbinia** *arrived unannounced and proceeded to race up and down the lines of anchored warships at an unheard of speed of 34.5 knots. The Royal Navy were unable to stop it as their own boats were too slow! The Admiralty were astounded with the unplanned display and after further trials with* **Turbinia** *placed an order for two turbine-powered torpedo boat destroyers – the turbines were manufactured by Parsons' new company at Wallsend. Launched in 1899, HMS* **Viper** *and HMS* **Cobra** *had a top speed of 36.5 knots and led the world in naval technology. By 1901 the world's first turbine-powered commercial ship, the TS* **King Edward** *had entered service on the River Clyde (see 'Doon the Watter'). A new age of marine propulsion had dawned and Britain was soon to launch the world's fastest and most powerful warship ever built, HMS* **Dreadnought** *(see 'Britannia Rules the Waves'). While Parsons died in 1931, his revolutionary* **Turbinia** *was eventually preserved and can be seen today at the Discovery Museum in Newcastle.*

Switch on the Lights
Generating electricity from steam

Designed by Thomas Edison, the first public electricity power station in Britain started operation in London in 1882. At its heart was a steam engine that mechanically drove a generator nicknamed 'Jumbo' – customers included the City Temple, Old Bailey and the General Post Office.

Following Charles Parsons' successful experiment with steam turbines (see 'A Demonstration That Shook the World') these were soon being used to generate electricity in thermal power stations – the high-pressure superheated steam raised in the boiler turned a multi-bladed turbine which in turn drove a generator. A plentiful supply of coal led to the building of hundreds of these power stations in Britain and even by the early 1950s, 90 per cent of all electricity generated in Britain was fired by coal, the remainder being oil-fired.

Steam turbine

BATTERSEA POWER STATION

Until the 1930s, British electricity was generated at small power stations owned by municipal companies. In an effort to standardise supplies through a National Grid several private power companies were formed, of which the London Power Company was one of the first. Their new coal-fired super power station, Battersea 'A', with coal delivered by a constant stream of Thames lighters, came on stream alongside the Thames in 1933 while the second station, Battersea

'B', was commissioned in 1953. 'A' was decommissioned in 1975 and 'B' in 1983. The iconic building – the largest brick building in Europe – with its Art Deco interior now faces an uncertain future and has been placed on the 'Buildings at Risk Register'. In addition to the use of coal, oil and, more recently, natural gas, the first nuclear-powered power station at Calder Hall came on stream in 1956. However even by the end of the 20th century around 30 per cent of Britain's electricity

was generated at coal-fired power stations, with gas at 40 per cent, oil at 1 per cent and nuclear at 20 per cent – there is still a long way to go before carbon-producing fossil fuels are phased out. While producing no carbon emissions, the nine nuclear power stations operating in the UK are nearly life-expired and will need to be replaced soon by a more modern design.

Edwardian Elegance
Harry Wainwright's long-lived locomotive designs

Harry Wainwright was born in Worcester in 1864 and followed in his father's footsteps when he was appointed Carriage & Wagon Superintendent at the Ashford Works of the South Eastern Railway in 1896. Three years later the SER joined the London, Chatham & Dover Railway to become the South Eastern & Chatham Railway and Wainwright was appointed Locomotive, Carriage & Wagon Superintendent of the new company.

Initially he had to oversee existing locomotive orders placed by his predecessors but his first designs, built by outside contractors and at Ashford, started to appear in 1900. Many of these elegant Edwardian locos, such as Wainwright's 'C' Class 0-6-0, 'E' Class 4-4-0 (rebuilt by Richard Maunsell as Class 'E1'), 'H' Class 0-4-4T, 'L' Class 4-4-0 and 'P' Class 0-6-0T, saw service through to the BR era in the early 1960s.

Wainwright retired in 1913 and died in 1925.

Kings of the Road
The story of steam lorries

The early years of the 20th century was a boom time for steam lorry manufacturers but this was soon cut short by the onslaught of the internal combustion engine.

Thornycroft
John Thornycroft was born in Rome in 1843 and went on to study engineering in Glasgow before establishing the Steam Carriage & Wagon Company at Chiswick, West London, in 1864. A year later the notorious Red Flag Act (see 'The Red Flag Act') halted the

development of steam road vehicles so Thornycroft switched to building steamships. His HMS *Lightning* (see 'John Thornycroft's Torpedo Boat') was the first naval vessel in the world to carry torpedoes.

Once the Red Flag Act had been repealed Thornycroft started producing steam vehicles again and with this business expanding moved production to Basingstoke in 1898. In 1899 the company supplied an experimental steam wagon to the Royal Engineers and built London's first double-decker steam bus that ran between Shepherds Bush and Oxford Circus. His shipbuilding business became a separate company (later to become Vosper Thornycroft) in 1904 when it moved to Southampton.

In 1901, Thornycroft won a Government competition to supply three-ton steam lorries to the Army. With a reputation for reliability, the company's steam lorry production continued until 1919, by which date the internal combustion engine had taken over as the favoured power unit. John Thornycroft died in 1928.

Edwin Foden & Sons
Edwin Foden was born in 1841 near Sandbach in Cheshire. After gaining an apprenticeship with a local agricultural engineering company, Platt & Hancock, he gained further experience at the Crewe Railway Works of the London & North Western Railway. He returned to Platt & Hancock in 1860 and later became a partner of the renamed company Hancock & Foden.

During the 1880s Foden designed and built massive stationary compound steam engines for industrial uses – examples of these can be seen today at the National Waterways Museum in Gloucester and at the Kidwelly Industrial Museum in South Wales. Foden also built his first steam tractor in 1882 and within a few years had become a major manufacturer of these vehicles for use as agricultural traction engines. In 1887, his partner George Hancock retired and the company was renamed Edwin Foden & Sons.

The repealing of the Red Flag Act (see 'The Red Flag Act') in 1896 led Foden to develop highly successful steam lorries, the first of which was produced with an overtype engine in 1900. Although coming second in a Government competition held in 1901 to supply steam lorries for the Army, Foden went on to produce a series of highly successful steam lorries until production stopped in 1935, by which date around 6,500 such vehicles had been built.

Although Edwin Foden died in 1911 the company of Edwin Foden & Sons remained intact until an acrimonious family split in 1932 when Edwin's younger son, E. R. Foden, left to found the diesel lorry company known as ERF.

Mann's Patent Steam Cart & Wagon Company

James Mann of Leeds started business in 1894 and his company, Mann & Charlesworth, went on to build stationary steam engines, traction engines and steam rollers. In 1898 he patented an agricultural steam cart which had a traction engine at the front and a heavy roller at the rear. So successful was this that in the following year he set up Mann's Patent Steam Cart & Wagon Company which, by 1901, was operating from new premises in Hunslet, Leeds.

At first highly successful, the company applied steam power to numerous road applications including road rolling, dustcarts, tar spraying, brewer's wagons and buses. However, following the end of World War I, the company struggled to survive against increasing competition from the internal combustion engine and, in 1929, it closed.

Sentinel

This major manufacturer of steam lorries (and later railway locomotives) started life in Glasgow as Alley &

Steam lorry

MacLellan in the 1870s. The later company name of Sentinel Waggon Works Ltd derived its title from the original Sentinel Works in the city. Initially building steam ships, the company started producing steam road vehicles in 1905 when it introduced a highly successful two-cylinder vertical-boiler steam wagon. Known as the Standard Sentinel it remained in production, albeit with modifications, until the introduction of the larger Super Sentinel steam lorry in 1923.

In 1915 a new company, Sentinel Waggon Works, was formed to manufacture steam lorries in Shrewsbury. First introduced in 1923, the larger Super Sentinel steam lorry was built in a new 'hi-tech' factory at Shrewsbury incorporating modern production line techniques. Always at the forefront of steam lorry technology the company went on to design a new steam lorry in 1934 incorporating a compact vertical-boilered four-cylinder underfloor engine that outsold all other British steam lorries – nearly 4,000 were built with different wheel configurations.

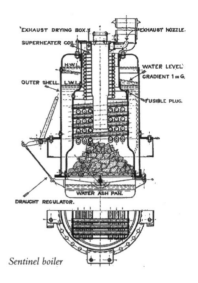

Sentinel boiler

However, the introduction of punitive road taxes for steam-powered road vehicles in the 1930s by the British Government, along with increasing competition from diesel-engined lorries, eventually spelt the end for companies such as Sentinel, although the last Sentinel steam lorries were exported to southern Argentina as late as 1950. The company continued to make diesel lorries, steam and diesel industrial railway locomotives until being taken over by Rolls-Royce in 1957.

Steam Road Vehicles Galore

Where to see steam road vehicles

There are numerous outdoor events held each year where preserved steam road vehicles and traction engines can be seen in action. For a full list visit: www.steamfair.co.uk.

The following events are a small selection:

SHREWSBURY STEAM ENGINE & VINTAGE VEHICLE RALLY

Onslow Park, Shrewsbury, Shropshire SY3 5EE
Tel: 01743 792 731
Website: www. shrewsburysteamrally.co.uk
Held over the August Bank Holiday weekend

THE GREAT DORSET STEAM FAIR

Child Okeford, Blandford, Dorset DT11 8HX
Tel: 01258 860361
Website: www.gdsf.co.uk
Held late August/early September

BEAULIEU NATIONAL MOTOR MUSEUM

Beaulieu, Brockenhurst, Hampshire S042 7ZN
Tel: 01590 612345
Website: www.beaulieu.co.uk
Steam Revival event held early June

ABBEY HILL STEAM RALLY

Yeovil Showground, Dorchester Road, Yeovil, Somerset
Tel: 01935 863199
Website: www.abbeyhillrally.co.uk
Held over May Bank Holiday weekend

BORDER CITY STEAM FAIR

Rickersby Park, Carlisle, Cumbria CA3 9AA
Tel: 07785 794 878
Website: www. bordercitysteamfair.co.uk
Held over Whitsun Bank Holiday weekend

SOMERSET STEAM SPECTACULAR

Low Ham, Nr Langport, Somerset
Tel: 01761 470867
Website: www. somersetspectacular.co.uk
Usually held mid-July

'Directors' and 'RODS'
John Robinson at Gorton

John Robinson was born in Bristol in 1856 and served an apprenticeship with the Great Western Railway at Swindon between 1872 and 1878. In that year he went to work for his father, a Locomotive Superintendent in Bristol, before joining the Waterford & Limerick Railway as Locomotive Superintendent in 1884. A prolific locomotive designer, Robinson produced 12 classes of loco for the Irish company over the next six years, some of them remaining in service on CIE until 1959. In 1900, Robinson moved back to England where he joined the Great Central Railway as Locomotive Superintendent at Gorton Works, near Manchester. Two years later he was appointed Chief Mechanical Engineer of the GCR, a post he held until the company became part of the newly-formed LNER in 1923.

Robinson was not only a prolific locomotive designer, producing 28 different classes for the GCR, but also a prolific inventor with 45

Class '04' heavy goods

patents to his name. Most notable of his locomotive types are his 4-4-0 'Directors' (later Class 'D10/D11') of which 45 were built, the majority staying in service until 1960, the six massive 0-8-4T shunting engines for Wath Marshalling Yard and the 2-8-0 heavy goods locos (later Class '04'). The latter was chosen as the standard freight engine by the Railway Operating Department during World War I and over 500 were built, many of them seeing service overseas during both World Wars. Robinson retired at the end of 1922 and died in 1943.

Early Standardisation
George Jackson Churchward's great legacy

Born in Stoke Gabriel, Devon, in 1857 George Churchward

started his apprenticeship in 1873 with the broad gauge South Devon Railway in Newton Abbot. He was transferred to the Great Western Railway's drawing office at Swindon Works in 1876 and went on to develop an automatic vacuum braking system before being promoted to Assistant Manager in the Carriage and Wagon Works in 1881. Churchward's rise to power at Swindon was fairly rapid: in 1885 he became Manager of the C&W Works; in 1895, Locomotive Works Manager; in 1897, Principal Assistant to the Chief Mechanical Engineer, William Dean (see 'The Death of the Broad Gauge'); and in 1902, Locomotive, Carriage and Wagon Superintendent – this latter 'holy' position had its title changed to Chief Mechanical Engineer in 1916.

Churchward's time at Swindon was marked by his advanced locomotive designs, the concept of standardisation of parts using interchangeable boilers and driving wheels, the development of ingenious valve gear and the application of superheating. He developed nine standard locomotive designs but probably his most famous were the 'City' Class 4-4-0s and the 'Star' Class 4-6-0s. His unique 'Pacific' loco, *The Great Bear*, was the first of its type in Britain but proved too heavy for much of the GWR network. Churchward retired in 1922 and continued to live in Swindon in a company-owned house until he was killed by an express train while crossing the main line in 1933.

The Great Bear

Made in France
The Great Western Railway's compound experiment

It was highly unusual for the GWR to look beyond Swindon, let alone abroad, for a new steam locomotive type but in 1903 the company's Locomotive, Carriage and Wagon Superintendent, G. J. Churchward, did just that. Wishing to compare the performance of his own two-cylinder 'simple' engines with the best of European designs over the demanding South Devon Banks, he ordered three 4-4-2 'Atlantic' de Glehn compound locomotives of the type that were then running on the Chemin de Fer du Nord in Northern France.

All built by Société Alsacienne de Constructions Mécaniques in Mulhouse, the first to be delivered, in 1903, was numbered 102 and appropriately named *La France*. The loco had two high-pressure cylinders fitted inside the frames driving the front pair of driving wheels, and two outside low-pressure cylinders driving the rear wheels. She was immediately put to work on principal GWR expresses such as the newly-inaugurated 'Cornish Riviera Express' but despite putting in good performances offered no significant improvement on Churchward's two-

cylinder 4-4-2 No. 171 *Albion* – the latter loco had been rebuilt from one of his prototype 'Saint' Class 4-6-0s to gain a direct comparison with the French loco. *La France*, along with the other two larger-boilered French locos, No. 103 *President* and No. 104 *Alliance* (both ordered in 1905), remained in service on the GWR (albeit with modifications) until 1926.

Although all three of the de Glehn locos performed well, at the end of the day Churchward still opted for his own 4-6-0 design of which the 'Saint' Class was the first example. However, the experiment wasn't a failure by any means as the de Glehn front bogie design was taken up by Churchward for use on his own locomotives and, later, by Swindon-trained William Stanier at the LMS. While at Swindon, Stanier described the fast-running engines as 'beautifully designed and an inspiration to all'.

GWR de Glehn Atlantic

Britannia Rules the Waves
HMS Dreadnought, *the world's first modern battleship*

HMS Dreadnought

Charles Parsons' revolutionary steam turbine (see 'A Demonstration That Shook the World') made the existing warships of the world's navies obsolete overnight and marked the start of an arms race that was to climax at World War I's Battle of Jutland in 1916.

Laid down at Portsmouth in 1905, HMS *Dreadnought* was a 527ft-long, armour-plated (up to 11 inches thick around the waterline), 18,120-ton battleship that marked a major milestone in naval technology. At her heart were two paired sets of Parsons' direct drive steam turbines driven by high-pressure steam (at 250psi) raised in 18 coal-fired boilers, giving the ship a top speed of 21 knots. Another world naval first, she mounted five pairs of 12-inch guns in turrets, a secondary armament of 27 3in guns, and five torpedo tubes. She had a full complement of around 750 crew, and carried nearly 3,000 tons of coal (plus over 1,000 tons of fuel oil to improve the burn rate of the coal)

that gave a range of around 7,500 miles before refuelling. Compared to other warships around the world she was unbeatable and was the pride of the British Royal Navy.

Dreadnought was launched by King Edward VII at Portsmouth early in 1906 and began her sea trials in record-breaking time, only one year after being laid down. She served as flagship of the Royal Navy until 1911 and was stationed at Scapa Flow during World War I, escaping the Battle of Jutland in 1916 due to a refit. She was decommissioned in 1919 and scrapped in 1923.

Well-Built and Long-Lived
The locomotives of William Reid

William Reid was born in Glasgow in 1854 and started

his long career with the North British Railway as an apprentice at Cowlairs Works in the city. He slowly rose through the ranks and was appointed Chief Mechanical Engineer on the retirement of Matthew Holmes (see 'A Cowlairs Man') in 1903, a position he held until 1919. He died in 1932.

During his 17 years at Cowlairs, Reid rebuilt many of Holmes' designs, introducing superheating for the first time on the NBR, as well as designing 12 new classes of locomotive. Although Holmes favoured the 4-4-0 wheel arrangement for most of his express passenger locomotives he also designed one class of the 4-4-2 'Atlantic' type which saw service on the East Coast Main Line north of Edinburgh and on the Waverley route to Carlisle. Most famous of his 4-4-0s are the Class 'D29', the 'Scott' Class and the 'D34' Glen Class, some of which survived on the West Highland Line until 1961. On the freight side, his most successful 0-6-0 was without doubt the Class 'J37', introduced in 1914, which saw service through to the end of steam in Scotland in 1967.

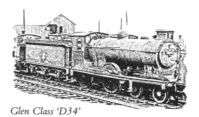

Glen Class 'D34'

The Cunard Twins
RMS Mauretania *and RMS* Lusitania

By the beginning of the 20th century the Germans were ruling the Atlantic waves with their fast and luxurious liners, SS *Kaiser Wilhelm der Grosse* and SS *Deutschland*, while the American-owned White Star Line was also making inroads into the important transatlantic business. Faced with this foreign competition on the doorstep the Cunard Line, with financial backing from the British Government, commissioned two new 'superliners'.

Although originally designed to be powered by large triple expansion steam engines, the development of Parsons' revolutionary steam turbine led to the latter form of propulsion being chosen for the Cunard twin ships. The first, RMS *Mauretania*, was laid down at Swan Hunter's shipyard in Wallsend, near Newcastle in 1904. Her sister ship, the RMS *Lusitania*, was laid down at John Brown's on Clydebank in the same year.

When launched in 1906, *Mauretania* was the largest ship in the world and by 1909 had also became the fastest ocean liner, crossing the Atlantic at an average speed of 26 knots – a record she held until 1929. Weighing in at nearly 32,000 tons, this luxury leviathan was 790ft long, powered by four Parsons' steam turbines with a power output of 70,000hp, and could carry 2,165 passengers with 800 crew. In service for 28 years, *Mauretania* exceeded all expectations for Cunard on the competitive transatlantic run and even served as a troopship and hospital ship during World War I. She was finally withdrawn in 1934 and scrapped

RMS Mauretania

RMS Lusitania

at Rosyth on the Firth of Forth.

Her sister ship, RMS *Lusitania*, was not so lucky. She entered service with Cunard in 1907 and made regular high-speed crossings of the Atlantic until, in 1915 during World War I, she was torpedoed by the German U-boat *U-20* off the coast of Ireland. She sank in 18 minutes, and of the 1,265 passengers (many of them well-known Americans) and 694 crew on board, only a total of 761 were saved – this tragic act certainly hastened America's later entry into the war.

CHARLES JOHN BOWEN-COOKE

Charles Bowen-Cooke was born in Huntingdon in 1859 and started his apprenticeship at the London & North Western Railway's Crewe Works in 1875 under the Works' Manager Francis Webb (see 'The

Giant of Crewe'). Primarily employed in the motive power department, Bowen-Cooke worked his way up the corporate ladder at Crewe to become Chief Mechanical Engineer in 1909. While CME he developed the use of superheating and was responsible for the design and construction of several express locomotive classes including 90 of the 'George the Fifth' Class 4-4-0s, 130 of the 'Claughton' Class 4-6-0s and the heavy freight 'G1' and 'G2' Class 0-8-0s. Bowen-Cooke died in office in 1920.

Steam Supremo
The North Eastern Empire of Vincent Raven

Vincent Raven was born in Great Fransham, Norfolk, in 1859. He started his railway career in 1877 as an apprentice with the North Eastern Railway and by 1893 had become Assistant Mechanical Engineer to Wilson Worsdell (see 'In His Brother's Footsteps'). Raven was promoted to Chief Mechanical Engineer when Worsdell retired in 1910 and, despite being an advocate

of main line electrification, went on to design nine classes of steam locomotives for the NER, many of which survived into the 1960s.

Favouring three-cylinder designs, Raven introduced his Class 'Z' 4-4-2 ('Atlantic') in 1911 and these were soon putting in fine performances on East Coast Main Line expresses between York and Newcastle. Heavy freight trains, in particular coal and iron ore, were the lifeblood of the NER and Raven designed two superb 0-8-0 classes, later to be classified 'Q6' and 'Q7', that numbered 135 locos – some members of the former class remained in service until the end of steam in the north-east in 1967. Similarly successful were his Class 'S3' 4-6-0s (later Class 'B16'). Raven's final design was the '2400' Class 4-6-2 (later Class 'A2'), of which five were built between 1922 and 1924.

Raven was knighted for his services during World War I in 1917, retired in 1922 and died in 1934.

The Only Survivor
The White Star Line's RMS Olympic

Designed to compete with the Cunard Line's new luxury transatlantic 'superliners' *Mauretania* and *Lusitania*, the American-owned White Star Line built three new liners designed to be the largest and most luxurious ships to sail the Atlantic. All were built by Harland & Wolff in Belfast and fitted with tried and tested reciprocating triple expansion steam engines for the two outboard propellers along with a low-pressure steam turbine for a centre propeller.

With slight variations, their statistics were pretty staggering: weighing in at around 46,000 tons each, the three ships were 882ft long, 175ft high with nine decks, and could carry around 2,400 passengers and 880 crew across the Atlantic at an average speed of 21 knots. Steam for their two four-cylinder reciprocating triple-expansion engines and a low-pressure turbine was generated in 29 boilers.

The first of these giants, RMS *Olympic*, was launched in 1910 – *Titanic* followed in 1911 and *Britannic* in 1914. For a short time the largest ocean-going liner in the world, *Olympic* made her maiden voyage to New York in 1911 but on her return was involved in a collision with a Royal Navy cruiser, putting her out of service until the next year. Following the *Titanic* disaster (see 'A Night to Remember') she was refitted with more lifeboats and improved watertight bulkheads before serving as a troopship in World War I, during which time

RMS Olympic

she sank a German U-boat by ramming it. After the war *Olympic* was refitted and converted to oil burning, subsequently remaining in service on the Atlantic run until she was retired in 1935. She was broken up in 1937.

THE REID-RAMSAY TURBINE-ELECTRIC

A power station on wheels, this strange-looking and complicated contraption was built by the North British Locomotive Company in 1910. With a wheel arrangement of 4-4-0+4-4-0 and weighing 132 tons it was fitted with a conventional steam boiler, with superheating, that drove an impulse turbine, which in turn drove a generator to produce electricity. The power produced then drove four traction motors fitted to the two pairs of driving wheels on each bogie. It was apparently tested by the Caledonian Railway and the North British Railway but the results were obviously not encouraging as the loco was never seen in action again.

From Steam to Fruit
Lawson Billinton's classic Brighton locos

The son of locomotive engineer Robert Billinton (see 'A Yorkshireman in Brighton'), Lawson Billinton was born in Brighton in 1882 and worked under his father as a draughtsman at the London, Brighton & South Coast Railway's Brighton Works and Locomotive Superintendent at New Cross. He became Locomotive Superintendent for the LBSCR at Brighton in 1911 where he was responsible for the design and building of the 'E2' Class 0-6-0T, the highly reliable 'K' Class 2-6-0 and the 'L' Class 4-6-4T. The latter were used for express passenger service with *Remembrance* being the last locomotive built for the LBSCR at Brighton before the Big Four Grouping and were later rebuilt as the 4-6-0 'N15X' Class by Richard Maunsell (see 'A Little Irish Magic'). Lawson Billington retired from the LBSCR on the formation of the Southern Railway in 1923 and became a fruit farmer until his death in 1954.

A Night to Remember
The ill-fated RMS Titanic

The second of three new 'superliners' built for the White Star Line, RMS *Titanic* was launched at Harland & Wolf's shipyard in Belfast in 1911. As history so famously records, the sister ship of RMS *Olympic* very soon ran out of luck, striking an iceberg during her maiden voyage to New York on 14 April 1912.

While *Titanic* carried only 20 lifeboats with a capacity for 1,178 people there were actually a total of 2,223 passengers and crew on board. Of these, 1,504 perished in the freezing waters or went down with the ship that sank just two hours and forty minutes after the collision. Help came only two hours and thirty minutes later when the Cunard liner RMS *Carpathia* arrived on the scene and rescued the survivors.

Much has been said and written about the *Titanic* disaster, one of the worst such incidents in peacetime history, and many questions remain unanswered. Where the blame lay and however the manner of her sinking,

though, the lack of lifeboats was still the main contributory factor to the high loss of life. Her sinking taught valuable lessons not only for the construction of future passenger liners but also for the provision of adequate life-saving equipment.

The wreck of the *Titanic* lies at a depth of around 12,500ft on the bed of the Atlantic and was first discovered by a joint French–American team using side-scan sonar in 1985. Later, remote-controlled submersibles and manned submersibles found that the ship was split in two on the ocean floor. Artefacts from the wreck were brought to the surface in the 1990s – a collection of these can be seen at the Merseyside Maritime Museum in Liverpool.

RMS Titanic

RMS Britannic

A SECOND SINKING

The third and final member of the White Star Line's 'superliner' class of transatlantic liners was RMS **Britannic**. *She was not launched until 1914 due to design alterations following the* **Titanic** *disaster (see 'A Night to Remember'). Never fulfilling her intended role as a transatlantic liner, she was used instead as a hospital ship during World War I until she was sunk by a mine off the coast of Greece in 1916. Fortunately, this time, help was close at hand and only 30 people perished whilst over 1,000 were saved.*

Underwater Steam
The 'K' Class submarines

While the development of slow diesel-electric submarines was already well underway, by World War I the British Admiralty was looking for faster vessels that could keep up with their fleet on the surface. Thus the 'K' Class submarine was born – designed in 1913, it was powered by two oil-fired Yarrow boilers, each powering a steam turbine that in turn drove two three-bladed propellers. Funnels from the boilers were folded down before the vessel submerged – one can only guess at the searing temperatures in the boiler room while the vessel was underwater! Weighing nearly 2,000 tons, this 339ft-long monster had a maximum surface speed of 24 knots (8 knots submerged), a crew of 59, and was fitted with four torpedo tubes at the fore and aft, plus two mounted on the deck.

WHAT A KALAMITY

The accident-prone 'K' Class submarines soon earned the nickname 'Kalamity' Class. Six were sunk in accidents, many of these collisions with other naval ships or between 'K' Class submarines themselves and involved great loss of life. Only one was ever involved in enemy action but the torpedo it fired at a German U-boat failed to detonate. A total of 17 of these submarines were built between 1916 and 1923 but their

'K' Class submarine

unreliability led to their early demise. The majority had been scrapped by 1926 although the final improved version, 'K26', survived until 1931 when she was also scrapped as she exceeded the size limits set out in the 1930 London Naval Treaty. So ended the Royal Navy's experiment with steam-driven submarines until the first of the nuclear submarines in 1960. In that year the UK's first nuclear-powered submarine, HMS Dreadnought, was launched. In comparison to the 'K' Class she was 73ft shorter but weighed 3,500 tons. She was powered by a Westinghouse S5W nuclear reactor driving two geared steam turbines, which in turn drove one propeller shaft. With a complement of 113 crew, she could travel at 28 knots under the surface and could remain submerged as long as food supplies held out.

Flight Deck
The flat top HMS Argus

The importance of air power had been amply demonstrated during World War I and before the war had even ended the British had built the world's first aircraft carrier.

In fact, HMS *Argus* started life as an Italian liner but was converted into a carrier before she was launched on the Clyde in 1917.

Weighing 14,680 tons and powered by four Parsons' steam turbines, she was an ungainly looking ship with a totally unobstructed flight deck. Although commissioned at the very end of the war she did eventually see active service during World War II when she not only ferried aircraft out to the Mediterranean but also provided air cover for the lifesaving Malta convoys. This grand old lady was scrapped in 1947.

HMS Argus

FROM COAL TO OIL

With the British-protected oilfields of Persia and Iraq coming on stream, oil had become more readily available as an alternative source of fuel to coal by World War I. Moreover, coal firing of ships was a labour intensive, expensive and dirty operation – during World War I vast amounts of coal had been taken by train to the far north of Scotland where it was loaded on to colliers to make the sea trip to the main Royal Navy base at Scapa Flow in the Orkneys before being manhandled into the warships' coal holds. The first oil-fired Royal Navy vessels were destroyers introduced in 1907. The first oil-fired capital ships built for the Royal Navy were the 'Queen Elizabeth' Class that were completed between 1915 and 1916 – by the early 1930s coal-fired capital ships had become a thing of the past. However, coal-burning support ships remained in service for the Royal Navy well into World War II. The larger ocean-going passenger liners were also converted to burn oil after World War I.

'Castles' and 'Kings'
Charles Collett's classic designs

Charles Collett was born in London in 1871 and his childhood was spent close to Paddington station. In 1893 he joined the Great Western Railway working as a draughtsman in the company's drawing office at Swindon before becoming a section leader in 1897. In 1898 he became Assistant to the Chief Draughtsman and two years later was promoted to Technical Inspector of the Works. Within a short time he had risen again to Assistant Manager, a post he held until 1912 when he was promoted to Manager. Still on the way up the GWR corporate ladder he became Deputy Chief Mechanical Engineer in 1919 and, following Churchward's retirement (see 'Early Standardisation'), Chief Mechanical Engineer in 1922.

Collett's time as CME at Swindon was the high point in GWR locomotive design – his 'Castle' and 'King' Class 4-6-0s must surely go down in railway history as some of the finest steam locomotives

'Castle' Class

ever built – a fact which the GWR's publicity department used to great effect. Collett retired in 1941 and died in 1952.

THE ARMSTRONG-WHITWORTH TURBINE-ELECTRIC LOCOMOTIVE

Following on from the North British Locomotive Company's experimental steam turbine electric locomotive of 1910 (see 'The Reid-Ramsay Turbine-Electric'), Armstrong-Whitworth of Newcastle built an even stranger-looking locomotive which was tested on the Lancashire & Yorkshire Railway in 1922. Weighing in at 156 tons, this loco had a wheel arrangement of 2-6+6-2 with each set of driving wheels being powered by two electric motors. The front unit carried the boiler, a compound turbine and a generator, while the rear unit carried coal, water and a massive air-cooled rotary evaporative condenser, the fan of which looked not unlike that of a jet engine. The contraption turned up at Horwich and was found to be so overweight that any main line testing was severely restricted. On the few runs that it made it proved inferior in performance to conventional steam locos and also returned very poor fuel consumption. By 1923 it had been returned to the manufacturers and was quietly scrapped.

The Horwich Crabs
George Hughes' finest hour

George Hughes was born in Cambridgeshire in 1865 and became a premium apprentice at the London & North Western Railway's Crewe Works in 1882. In 1887, he went to work as a fitter and erector of steam locomotives at the Lancashire & Yorkshire Railway's new works in Horwich. Over the next 18 years Hughes slowly worked his way up the L&YR's promotional ladder: in 1888 he was in charge of testing at Horwich; in 1894 he was running the Horwich gas works;

'Crab' Class

in 1895 he was promoted to Chief Assistant in the Carriage & Wagon Department at Newton Heath; in 1899 he became Works Manager at Horwich and Principal Assistant to Henry Hoy (see 'Could Try Harder'), finally succeeding him as Chief Mechanical Engineer in 1904.

While at Horwich, Hughes was particularly interested in the development of heavy goods locos and experimented with high-temperature superheating and compounding. His locomotive designs for the L&YR include the four-cylinder 4-6-0s, nicknamed 'Dreadnoughts', and their subsequent development into powerful 4-6-4Ts that were actually built under London, Midland & Scottish Railway management in 1924. Of course, Hughes' greatest contribution to the development of steam locomotives were his 245 distinctive 'Crab' 2-6-0s, built by the LMS at Horwich and

Crewe between 1926 and 1932. As a highly successful, mixed-traffic loco, many of this class remained in active service for BR in England and Scotland until 1967.

Hughes remained as CME of the L&YR until 1922. In that year the L&YR amalgamated with the LNWR and he became the CME of the expanded LNWR company. Following the Big Four Grouping of 1923 he became CME of the London, Midland & Scottish Railway but he retired to Norfolk only two years later. A cheerful man to the end, he died in Stamford in 1945.

Dependable, But Uninspiring
The workhorses of Henry Fowler

Henry Fowler was born in Evesham in 1870 and, after studying metallurgy in Birmingham, served an apprenticeship at the Lancashire & Yorkshire Railway's Horwich Works. From 1891 to 1895 he worked in the L&YR's testing department under George Hughes (see 'The Horwich Crabs'), finally becoming the Head of Department, followed by five more years as Gas Engineer for the company. Fowler joined the Midland Railway in Derby in 1900 as Gas Engineer & Chief of the Testing Department – he was promoted to Assistant Works Manager in 1905, Works Manager in 1907 and Chief Mechanical Engineer in 1909.

Following his secondment as Director of Production in the Ministry of Munitions during World War I for which he was knighted, Fowler was appointed Deputy CME (under George Hughes) of the newly-formed London, Midland & Scottish Railway in 1923. Hughes retired in 1925 and Fowler took over as Chief Mechanical Engineer of the LMS, a position he held until his retirement in 1933. Sir Henry Fowler died in 1938.

During his time as CME of the LMS, Fowler was responsible for the design of many types of locomotive, built in their hundreds and the majority of which (albeit with some rebuilding by William Stanier) remained in service into the BR era

'Royal Scot' Class

and the early 1960s. They include the Class '6P' 'Patriot' and '7P' 'Royal Scot' 4-6-0s, Class '3MT' 2-6-2Ts, Class '3F' 'Jinty' 0-6-0Ts, Class '2P' 4-4-0s, Class '4MT' 2-6-4Ts, Class '4F' 0-6-0s and Class '7F' 0-8-0s. He even managed to escape from the small-engine policy of Derby with his 2-6-0+0-6-2 Garratts, and his Somerset & Dorset Joint Railway '7F' 2-8-0s.

HENRY FOWLER'S HIGH-PRESSURE 'FURY'

While Chief Mechanical Engineer of the London, Midland & Scottish Railway, Henry Fowler designed an experimental high-pressure locomotive. Named **Fury** *and numbered 6399, it was based on the frames of his 'Royal Scot' Class 4-6-0 and was built by the North British Locomotive Company in 1929. The engine was a three-cylinder compound, a small high-pressure* *cylinder being situated between the frames and two low-pressure cylinders outside the frames. It was fitted with a Schmidt high-pressure water-tube boiler which consisted of three distinct systems: an ultra high-pressure closed-circuit system with a pressure of 1,400–1,800psi which transferred heat from the firebox to the second boiler; a second boiler that provided steam at 900psi for the high-pressure cylinder; and a third, low-pressure boiler operating at 250psi that drove the two low-pressure cylinders. During trials in 1930, when the loco was based at Glasgow Polmadie shed, one of the ultra high-pressure boiler tubes burst, killing a technician. Although the trials continued until February 1933 the results were not encouraging and the loco was rebuilt in 1935 as a conventional 'Royal Scot' and given the name* British Legion.

A Powerful Man
Robert Urie's 'big' engines

Robert Urie was born in Ayrshire in 1854 and, after serving apprenticeships at several Scottish locomotive manufacturers, went on to become Chief Draughtsman and then Works Manager at the Caledonian Railway's St Rollox Works. In 1897, he was appointed by Dugald Drummond (see 'Locomotive Longevity') as Works Manager at the London & South Western Railway's Nine Elms Works, moving to the new works at Eastleigh in 1909. He was appointed Chief Mechanical Engineer of the LSWR following Drummond's death in 1912.

By his retirement in 1923, Urie had introduced three classes of powerful 4-6-0 for main line work – his 'H15', 'N15' ('King Arthur' Class) and 'S15' all served through to the BR era in the early 1960s, as did his powerful 'G16' Class 4-8-0T and 'H16' 4-6-2T. Urie retired to his native Scotland in 1923, and died in 1937.

A Little Irish Magic
Richard Maunsell's reign at Ashford and Eastleigh

Richard Maunsell was born in Ireland in 1868 and went on to start an apprenticeship under

Henry Ivatt (see 'King of the Klondykes') at the Great Southern & Western Railway's Inchicore Works in Dublin in 1886. He later moved to England where he became a draughtsman for the Lancashire & Yorkshire Railway at Horwich Works, followed by a period as a locomotive foreman. In 1894, Maunsell left England to work as Assistant Locomotive Superintendent of the East India Railway but returned to Ireland two years later to take up the position of Works Manager at Inchicore, finally becoming Locomotive Superintendent of the GS&WR in 1911.

In 1913, Maunsell was appointed as Locomotive Superintendent of the South Eastern & Chatham Railway at Ashford Works. Over the next ten years he designed six classes of steam locomotive that were all built at Ashford. These included his 'N' and 'N1' Class 2-6-0s that gave sterling service well into the BR era, some surviving until 1966.

In 1923, the SE&CR became part of the new Southern Railway and Maunsell was appointed Chief Mechanical Engineer of this enlarged company, remaining in office until his retirement in 1937. As CME of the Southern, Mansell not only oversaw the expansion of third-rail electrification but was also responsible for the introduction of eight classes of steam locomotive, all of which saw service into the 1960s. Of note are his 4-4-0 'Schools' Class and the 4-6-0 'Lord Nelson' Class which, when introduced in 1926, were the most powerful 4-6-0s in Britain. He retired in 1937 and died in 1944.

Articulated Steam Locomotives
The story of Beyer-Peacock

'Schools' Class

Founded in 1853, the Manchester-based locomotive builders of Beyer-Peacock had supplied articulated steam locomotives to railways around the world since 1909. However, only two British railway companies bought examples of these powerful locos, the first being the LNER with its unique Class 'U1' 2-8-0+0-8-2 that was introduced in 1925. This 87ft-long monster – the most powerful steam loco in Britain – spent most of its life banking heavy coal trains over the heavily-graded Woodhead route and was withdrawn in 1955.

The lack of powerful goods engines on the LMS led the company to purchase 33 of the 2-6-0+0-6-2 Beyer-Garratts between 1927 and 1930. Primarily employed hauling heavy coal trains between the Nottingham coalfields and London, many were eventually fitted with revolving coal bunkers and remained in service until the 1950s.

Today, preserved narrow gauge Beyer-Garratt locomotives from South Africa can be seen in action on the Welsh Highland Railway between Caernarfon and Porthmadog in North Wales.

Small But Powerful
The world of miniature steam locomotives

The world of miniature (or 'minimum') railways owes much to Eton and Cambridge-educated Sir Arthur Heywood who went on to design novel steam locomotives for his private 15in-gauge railway that he built in the grounds of his country house near Duffield, Derbyshire, in 1874. Known as the Duffield Bank Railway, the mile-long line was built with sharp curves and tunnels, demonstrating Heywood's belief in the use of a basic, cheaply constructed railway for industrial and military uses. The Duke of Westminster took up Heywood's ideas and built a working 15in-gauge railway in the grounds of Eaton Hall, Cheshire, in 1896.

The early 20th century saw the building of miniature passenger-carrying railways at various seaside resorts around Britain. Probably the most famous of these is the Rhyl Miniature Railway that opened in 1911 and was initially operated by a 4-4-2 built by Wenman Bassett-

Lowke (1877–1953). This railway's success led to the ordering of six 'Atlantic' 4-4-2s from local builder Albert Barnes in 1920 – three of these magnificent machines are still in operation on the line today.

Sir Arthur Heywood died in 1916 and much of his railway equipment was bought by Bassett-Lowke for use on the recently purchased 3ft-gauge Ravenglass & Eskdale railway in Cumbria. Bassett-Lowke converted the 7-mile line to 15in gauge, using it to transport granite and tourists – it still fulfils the latter purpose today. The first locomotive on this line was the 4-4-2 *Sans Pareil* and the connection with Heywood still exists today, with parts of his 0-8-0 locomotive *Muriel* being incorporated in the R&ER's 0-8-2 *River Irt* of 1927.

By far the most famous designer of miniature steam locomotives

Henry Greenly's Hercules

was Henry Greenly (1876–1947) who went into partnership with Bassett-Lowke in 1909. He is best known for his superb one-third scale 'Pacific' locos, built by Davey Paxman of Colchester for the 15in-gauge Romney, Hythe & Dymchurch Railway in Kent. This eight-mile line was opened between Hythe and New Romney in 1927 and extended by a further five and a half miles to Dungeness a year later. The first two locomotives built for the line, *Green Goddess* and *Northern Chief*, along with eight others (including two 4-8-2s), are still in service today.

Small is Beautiful
Where to see miniature railways

FAIRBOURNE RAILWAY
**Beach Road, Fairbourne,
Gwnedd LL38 2EX
Tel: 01341 250362
Website: www.fairbournerailway.
com**
Length: 2 miles
Gauge: 12¼in

PERRYGROVE RAILWAY
Perrygrove Road, Coleford,
Gloucestershire GL16 8QB
Tel: 01594 834991
Website: www.perrygrove.co.uk
Length: 1 miles
Gauge: 15in

RAVENGLASS & ESKDALE RAILWAY
Ravenglass, Cumbria LA9 4QD
Tel: 01229 717171
Website: www.ravenglass-railway.
co.uk
Length: 7 miles
Gauge: 15in

RHYL MINIATURE RAILWAY
Central Station, Marine Lake,
Wellington Road, Rhyl LL18 1LN
Tel: 01352 759109
Website: www.
rhylminiaturerailway.co.uk
Length: 1 mile
Gauge: 15in

ROMNEY, HYTHE & DYMCHURCH
RAILWAY
New Romney Station, New
Romney, Kent TN28 8PL
Tel: 01797 362353
Website: www.rhdr.org.uk

Length: 13 miles
Gauge: 15in

WELLS & WALSINGHAM LIGHT
RAILWAY
Wells-Next-the-Sea, Norfolk
NR23 1QB
Tel: 01328 711630
Website: www.
wellswalsinghamrailway.co.uk
Length: 4 miles
Gauge: 10¼in

Collett v. Gresley
The 1925 Locomotive Exchanges

Held between 27 April and 2 May 1925, the famous locomotive trials between a GWR 'Castle' Class 4-6-0 and an LNER 'A1' 4-6-2 proved beyond doubt that Swindon's practice of building locomotives that operated at higher pressures along with long-travel valve motions that allowed full regulator and short cut-off working was a winning formula. A virtually noiseless GWR engine running at speed was quite simply an advertisement for its own efficiency.

The engines in question at the trials were Charles Collett's No.

4079 *Pendennis Castle* and Nigel Gresley's No. 2545 *Victor Wild*, the former being the shorter and lighter of the two by 12 tons. The trials were held between King's Cross, Grantham and Doncaster with each loco hauling an identical load. Not only did *Pendennis Castle* perform better out of King's Cross through the tunnels and up Holloway Bank than the 'A1' but also was more economical in both coal (using 6lbs less per mile) and water. *Pendennis Castle* proved the undisputed winner.

Withdrawn in 1964, *Pendennis Castle* has since been preserved and is currently being restored at the Great Western Society's Didcot Railway Centre in Oxfordshire.

NIGHT MAIL
Made by the GPO Film Unit in 1936, this classic 25-minute documentary film starred parallel-boilered 'Royal Scot' Class 4-6-0 No. 6115 Scots Guardsman. *The film was scripted by the poet W. H. Auden, narrated by John Grierson and featured music by Benjamin Britten. Its opening lines go down in the annals of railway history:*

This is the Night Mail crossing the border
Bringing the cheque and postal order...
Scots Guardsman *(in its rebuilt form with tapered-boiler) is one of only two 'Royal Scot' 4-6-0s that have been preserved – the other is No. 46100* Royal Scot.

The World Beater
Nigel Gresley's iconic 'Pacifics'

Nigel Gresley was born in Edinburgh in 1876 and was brought up in Derbyshire. Educated at Marlborough, he was apprenticed at the London & North Western Railway's Crewe Works before moving on to continue his training under John Aspinall (see 'A True Lancastrian') at the Lancashire & Yorkshire Railway's Horwich Works. After three years as Assistant Superintendent in the L&YR Carriage & Wagon Department at Newton Heath, he was appointed Superintendent of the Carriage & Wagon Department of the Great Northern Railway in 1905. Six years later, he succeeded H. A.

Ivatt (see 'King of the Klondykes') as Locomotive Engineer of the GNR. In 1923 the GNR became part of the newly formed London & North Eastern Railway and Gresley was appointed Chief Mechanical Engineer, a position he held at the LNER's Doncaster Works until his death in 1941.

Without a doubt he is best remembered for his streamlined 'A4' 'Pacifics', especially No. 4468 *Mallard* which still holds the world speed record for a steam locomotive which it achieved in 1938 with a top speed of 126mph down Stoke Bank, north of Peterborough. The record-breaking run nearly ended in disaster as the locomotive limped into Peterborough with an overheated big end and was unable to continue its journey to King's Cross. Now preserved, *Mallard* can be seen at the National Railway Museum in Shildon.

GRESLEY'S HIGH-PRESSURE 'HUSH-HUSH' LOCO

Built in great secrecy, Nigel Gresley's 4-6-4 four-cylinder high-pressure compound locomotive No. 10000 was fitted with a marine water tube boiler that operated at 450psi. Known as the 'hush-hush' locomotive, it was built at Doncaster and made its first trial run with a unique streamlined casing in 1929. Although designed to be fuel-efficient, its boiler never lived up to expectations and despite numerous modifications the loco made its last journey to Doncaster in 1936. Here, it was rebuilt with a conventional boiler and remained in service as BR No. 60700 until withdrawal in 1959. During his time at both the GNR and LNER, Gresley was responsible for 26 different locomotive types including the 'A1' 4-6-2 in 1922, the 'A3' 4-6-2 in 1927, and the 'V2' 2-6-2 in 1936. However, his most famous locomotive must surely be the beautifully streamlined 'A4' 4-6-2, which first appeared in 1935 and held sway on the East Coast Main Line until the early 1960s.

RMS Queen Mary

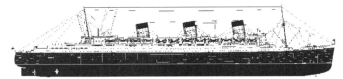

The Cunard Queens, Part 1
RMS Queen Mary

By 1930, competition from new German and French liners on the transatlantic route had become intense. The German liners *Bremen* and *Europa*, both weighing in at around 50,000 tons and powered by steam turbines, had just entered service and, not wishing to be left behind, Cunard planned two new 'superliners'. The first, later to be named *Queen Mary,* was laid down at John Brown's shipyard on Clydebank at the end of 1930. Her main rival, France's flagship, the 80,000-ton steam turbo-electric *Normandie*, was laid down just one month later.

A similar size to the *Normandie* but much larger than the German liners, *Queen Mary* soon hit problems on Clydebank when work was stopped on her at the end of 1931. A victim of the Great Depression, her future was in doubt until the Government stepped in with a loan, on condition that Cunard merged with the ailing White Star Line. Once the merger was completed work restarted on the ship and *Queen Mary* launched the ship on the Clyde in September 1934. Apparently Cunard originally wanted to name the ship after Britain's 'greatest queen', Victoria, but George V got the wrong end of the stick and thought the company was referring to his wife!

Queen of the Atlantic
The ultimate in art deco

Designed to carry over 2,000 passengers in style and comfort, the *Queen Mary*'s opulent Art Deco interior was designed by the Bromsgrove Guild of Fine Arts, a group of artists that were closely associated with the influential Arts and Crafts Movement. A mural in the First Class Dining room featured a map of the Atlantic with a moving model of the ship marking its position during the voyage.

With a crew of 1,100 and weighing in at just over 81,000 tons with a length of 1,019ft, *Queen Mary* was powered by four sets of Parsons' steam turbines driving four propeller shafts. Twenty-four oil-fired boilers raised steam for the turbines and the ship had a service speed of 28.5 knots. Her maiden voyage from Southampton to New York was in May 1936 and she was soon competing with *Normandie* for the fastest transatlantic crossing, known as the Blue Riband. She trounced the French ship twice, in August 1936 and again in August 1938 – the latter voyage took just 3 days, 21 hours at an average speed of 30.99 knots, a record she held until 1952 when the smaller SS *United States* took the lead.

World War II put an end to the transatlantic races and *Queen Mary,* along with her new sister ship *Queen Elizabeth* (see 'The Cunard Queens,

Part 2'), was despatched to Australia where she was refitted as a troop ship. Capable of carrying up to 20,000 troops at high speed she proved her worth many times over during the war, especially after the US entered the conflict when she transported hundreds of thousands of US troops across the Atlantic to Europe.

Following the end of the war *Queen Mary* was refitted with improved passenger accommodation and remained in service on the Southampton to New York run until 1967, by which time transatlantic jets had usurped most of the business. Put up for sale, she was bought by the city of Long Beach in California as a tourist attraction and converted into a floating hotel, restaurant and museum – she is even said to be haunted.

By comparison her modern namesake, the Cunard flagship *Queen Mary 2*, weighs in at 151,000 tons and, as a cruise ship, carries just over 3,000 passengers in total luxury supported by a crew of 1,250. The largest ocean liner ever built, she is powered by four of the latest electric propulsion pods.

In Graceful Retirement
Where to see Britain's historic steamships

ORIGINAL STEAM ENGINE FROM HENRY BELL'S COMET
Science Museum, Exhibition Road, London SW7 2DD
Tel: 0870 870 4868
Website: www.sciencemuseum.org.uk

ORIGINAL STEAM TURBINE FROM TS KING EDWARD
Riverside Museum, 100 Pointhouse Place, Glasgow G3 8RS
Tel: 0141 287 2660
Website: www.glasgowlife.org.uk/museums

PS WAVERLEY
For information on cruises aboard this ship:
Waverley Excursions Ltd, 36 Lancefield Quay, Glasgow G3 8HA
Tel: 0845 130 4647
Website: www.waverleyexcursions.co.uk

SIR WALTER SCOTT
Trossachs Pier, Loch Katrine, By Callander, Stirling FK17 8HZ
Tel: 01877 332000
Website: www.lochkatrine.com

PS MAID OF THE LOCH
Loch Lomond Steamship Company, The Pier, Pier Road, Balloch G83 8QX
Tel: 01389 711865
Website: www.maidoftheloch.com

NATIONAL TRUST'S GONDOLA
Coniston Pier, Lake Road, Coniston, Cumbria LA21 8AN
Tel: 01539 432733
Website: www.nationaltrust.org.uk/main/w-gondola

CLYDE PUFFER VIC 72
The Pier, Inverary PA32 8UY
Tel: 01499 302 213
Website: www.inverarypier.com

HMS WARRIOR
Main Road, HM Naval Base, Portsmouth PO1 3QX
Tel: 023 9277 8600
Website: www.hmswarrior.org

SS GREAT BRITAIN

Great Western Dockyard, Gas
Ferry Road, Bristol BS1 6TY
Tel: 0117 926 0680
Website: www.ssgreatbritain.org

TURBINIA

Discovery Museum, Blandford
Square, Newcastle-upon-Tyne,
Tyne & Wear NE1 4JA
Tel: 0191 232 6789
Website: www.twmuseums.org.
uk/discovery

HMS BELFAST

Morgan's Lane, Tooley Street,
City of London SE1 2JH
Tel: 020 7940 6300
Website: www.hmsbelfast.iwm.
org.uk

RMS QUEEN MARY

1126 Queens Highway, Long
Beach, California, CA 90802,
USA
Tel: 001 877 342 0738
Website: www.queenmary.com

RY BRITANNIA

Ocean Terminal, Leith,
Edinburgh EH6 6JJ

Tel: 0131 555 5566
Website: www.
royalyachtbritannia.co.uk

CALL BOY JUST KEEPS ON GOING

In 1930 the LNER 4-6-2, No. 2795
Call Boy, did 28 days continuous
running between Edinburgh
(Waverley) and London (Kings
Cross), a distance of 392 miles.
On the 24 weekdays of this period
it hauled the Flying Scotsman
non-stop between London and
Edinburgh, and on the four Sundays
it hauled a day express that stopped
intermediately. This engine therefore
covered 11,000 miles in four weeks
— 9,400 miles of these were made up
of the world's record daily non-stop
run. The mileage that could then be
covered by modern locomotives was
frequently surprising, and there is no
doubt that the engines of the period
were worked extremely hard, many
being double-crewed.

A Swindon Man at Crewe

William Stanier's outstanding contribution to locomotive design

Streamlined 'Coronation' locomotive

William Stanier was born in Swindon in 1876. His father was Chief Clerk to William Dean (see 'The Death of the Broad Gauge') at Swindon Works and, in 1892, the young William joined the Great Western Railway as an apprentice draughtsman. Stanier then worked his way up the GWR promotional ladder, being appointed Assistant Works Manager by George Churchward in 1912 before becoming Principal Assistant to Charles Collett (see "Castles' and 'Kings") in 1922. His time spent under both Churchward and Collett was to have a great influence on Stanier's later work for the London, Midland & Scottish Railway.

Stanier was finally headhunted by the LMS in 1932 following the retirement of Sir Henry Fowler (see 'Dependable, But Uninspiring') as the company's Chief Mechanical Engineer. Stanier quickly set about reorganising and standardising the company's mixed bag of locomotive designs, rapidly introducing some very successful locomotive classes that were to form the backbone of motive power for the LMS and BR until the end of steam in 1968. In addition to rebuilding Fowler's 'Royal Scot' and 'Patriot' Classes with taper-boilers, Stanier's prolific output included 841 of the 'Black 5' 4-6-0s, 190 of the 'Jubilee' 4-6-0s, and 776 of the '8F' 2-8-0s. However, his most famous locomotive designs were the powerful 'Princess Royal' and 'Coronation' 'Pacifics' that were the mainstay of West Coast Main Line expresses for 30 years.

William Stanier was knighted in 1943 and retired in 1944. He died in 1965.

THE END OF ROAD STEAM

Ernest Marples (and his cohort Dr Beeching) may be a hated figure for lovers of Britain's steam railways but the Conservative Minister for Transport in 1934, Oliver Stanley, caused just as much furore when he introduced swingeing new road taxes for steam road vehicles. This increased road tax for steam road vehicles to £100 per annum. Stanley also encouraged the use of the internal combustion engine by reducing tax on imported oil. Manufacturers of steam lorries such as Foden and Sentinel were the hardest hit but so too were the hard-pressed road-haulage industry and the coal miners who sold nearly one million tons of coal to the road transport business each year. Unable to pay the higher costs, most road hauliers scrapped their steam vehicles and moved over to diesel-engined vehicles instead.

The Great Innovator
The story of Oliver Bulleid

Oliver Bulleid was born in Invercargill, New Zealand, in 1882 but returned to Britain with his mother after the death of his father in 1889. In 1900, he began his apprenticeship with the Great Northern Railway at Doncaster Works and, in 1904, became Locomotive Running Superintendent and then, in 1905, Manager of the Works. Bulleid moved to Paris in 1908 where he worked for the Westinghouse Electric Corporation in its brake and signal division. Returning to England in 1910, he worked for the Board of Trade until 1912 when he was appointed assistant to the GNR's Chief Mechanical Officer, Nigel Gresley (see 'The World Beater'), a post he also held in the newly formed London & North Eastern Railway from 1923. In this role he was involved in the

'Merchant Navy' locomotive

development of the Class 'U1' 2-8-0+0-8-2 Garratt, and the Class 'P1' and 'P2' 2-8-0 locomotives.

Bulleid's big break came in 1937 when he was appointed CME of the Southern Railway following the retirement of Richard Maunsell (see 'A Little Irish Magic'). During his tenure as CME of the Southern, Bulleid designed and produced many groundbreaking steam locomotive types such as the innovative 'Merchant Navy' and 'West Country'/'Battle of Britain' 'Pacific' locos, the utilitarian Class 'Q1' 0-6-0 freight loco and the less successful 'Leader' Class with its steam-powered bogies. In 1949, Bulleid was appointed CME of Coras Iompair Eireann (Irish Railways) where he designed and built his famous turf-burning locomotive.

He retired in 1958 and died, in Malta, in 1970.

'OH! MR PORTER'

This classic comedy film was shot on the closed Basingstoke & Alton Light Railway in 1937 and featured Will Hay, Moore Marriott and Graham Moffatt. However the star of the film was Kent & East Sussex Railway's

2-4-0T Northiam (renamed Gladstone for the film) which was built by Hawthorn Leslie in 1899. After filming, the loco returned to work on the K&ESR, remaining in service until 1941.

The Cunard Queens, Part 2
RMS Queen Elizabeth

The larger sister ship to *Queen Mary, Queen Elizabeth* was laid down at John Brown's shipyard on the Clyde in 1936 and launched two years later. When completed she was the largest passenger liner ever built and carried this distinction for over half a century. Designed to carry 2,238 passengers and 1,000 crew, she weighed in at 83,673 tons and her four Parsons' oil-fired steam turbines driving four propellers had a combined power output of 160,000hp.

Completed at the beginning of World War II, she made her maiden voyage in some secrecy painted in battleship grey, sailing directly to New York in March 1940. Here she joined *Queen Mary* and *Normandie* before sailing on to Singapore where she was fitted out as a troopship. In this guise she carried over 750,000 troops from Australia and, from 1942, from North America, to the theatres of war in Europe, North Africa and Asia – her high speed allowed her to travel independently of convoys and to escape U-boat attacks.

Following the end of the war *Queen Elizabeth* returned to John Brown's shipyard on the Clyde where she was finally fitted out as a luxury ocean liner. She made her first commercial crossing of the Atlantic in October 1946 and, along with her sister ship *Queen Mary*, dominated the Southampton to New York service until the 1960s.

Sadly, unlike her sister ship, *Queen Elizabeth* came to an inglorious end. She was sold in 1968 as a tourist attraction at Port Everglades in Florida but this business venture soon ended in failure. She was subsequently sold to a Hong Kong businessman who planned to use her as a floating university. During her refit she mysteriously caught fire and sank in Hong Kong's Victoria Harbour – her remains lie

RMS Queen Elizabeth

on the harbour bottom to this day.

Her modern namesake, the 70,327-ton *Queen Elizabeth 2* took over as Cunard's flagship in 1969 – when built she was fitted with oil-fired steam turbines but these were replaced by a diesel-electric power plant in 1987. Now retired from her role as cruise ship, she currently resides in Dubai facing an uncertain future.

LMS *TURBOMOTIVE*

Built in 1935 at Crewe Works, the LMS 4-6-2, No. 6202 **Turbomotive,** *was a modified version of William Stanier's 'Princess Royal' Class locomotive, fitted with a turbine located ahead of the coupled driving wheels instead of cylinders. It was also fitted with a reverse turbine for going backwards. Considered to be an unqualified success due to its high* *thermal efficiency, this loco covered around 300,000 miles until a turbine failed in 1949. It was then rebuilt as a conventional loco, returning to service in 1952 as No. 46202* **Princess Anne** *– only a few months later the loco was involved in the tragic Harrow & Wealdstone accident when it was damaged beyond repair.*

A Mixed Bag
The ups and downs of Edward Thompson

Edward Thompson was born in Marlborough, where his father was a school master, in 1881. Like Nigel Gresley (see 'The World Beater'), Thompson was also educated at Marlborough School, but then went on to study mechanical science at Cambridge, gaining a BA Honours in the subject in 1902. Following this he was apprenticed at the Lancashire & Yorkshire Railway's Horwich Works before being appointed Assistant Divisional Locomotive Superintendent on the North Eastern Railway. During World War I he served in the Royal Engineers and was mentioned in despatches, subsequently joining the Great Northern Railway in 1919 to become Carriage & Wagon Superintendent at Doncaster. He held this post until 1930 when he was promoted to Workshop Manager at Stratford Works.

Nigel Gresley's sudden death in 1941 was unexpected and catapulted Thompson into the top job at the LNER where he was Chief Mechanical Engineer until his retirement to Westgate-on-Sea in 1946. He died in 1954.

The high point of Thompson's six years at Doncaster must surely be his two-cylinder Class 'B1' 4-6-0s, of which 410 were built between 1942 and 1952. They were highly successful performers although their life was cut short following the introduction of diesels in the early 1960s – the

'B1' Class

last examples were withdrawn in 1967 and two have been preserved. On the downside, many other Thompson locos were rebuilds of Gresley's engines – the 'A1/1' 4-6-2s (a rebuild of Gresley's first 'A1' 'Pacifics'), 'A2/1' 4-6-2s (enlarged Gresley 'V2' 2-6-2s), and 'A2/2' (rebuilds of Gresley's 'P2' 2-8-2s). Thompson's 15 'A2/3' Class 4-6-2s were built from new with the last examples being withdrawn in 1965.

THE OIL BURNERS
With vast deposits of indigenous coal available, Britain's railways had a seemingly endless supply of fuel to fire their steam locomotives. However, during times of industrial unrest or Arctic weather conditions the supply chain could seize up and in such cases some of Britain's railways temporarily converted some of their locos to oil burning. Prolonged miners' strikes before and after World War I, and the big freeze during the winter of 1947, saw many locomotives thus converted. During the latter period the GWR at Swindon converted examples of 2-8-0 freight locos along with a few 'Halls'

and 'Castles', but the experiment soon ended when it turned out that near-bankrupt Britain couldn't afford to buy the vast amounts of imported oil required to power them.

The End of a Glorious Era
Frederick Hawksworth at Swindon

Frederick Hawksworth was born in Swindon in 1884 and joined the Great Western Railway as an apprentice at the age of 14 years. A company man through and through, he slowly worked his way up the GWR hierarchy and was appointed Chief Locomotive Draughtsman at the time Charles Collett's (see "Castles' and 'Kings'") 'King' Class 4-6-0s were introduced in 1927. Hawksworth was appointed as Chief Mechanical Engineer of the GWR following Collett's retirement in 1941.

Despite wartime shortages he assembled a team of 100 per cent GWR men and by 1944 had introduced the 'Modified Hall' Class 4-6-0s which were a highly successful development of Collett's

'County' Class

'Hall' Class. Plans for a powerful 'Pacific' loco never got off the drawing board but the period 1945–47 was busy at Swindon with the introduction of Hawksworth's 'County' Class 4-6-0s and improved 'Castles' with four-row superheaters. Despite being a two-cylinder engine, the 'County' class was certainly a break away from GWR tradition with its high-pressure Stanier '8F' boiler and 6ft 3in driving wheels.

Other loco types introduced under Hawksworth were the '1600' Class and '9400' Class 0-6-0PTs, the latter with a taper boiler, as well as the '1500' Class 0-6-0PTs which were the only GWR locos ever fitted with Walschaerts valve gear. Following Nationalisation in 1948, Hawksworth remained in control at Swindon but was now responsible to the new Railway Executive. He retired in 1949 and died at the age of 92 years in 1976.

Fleet Air Arm
Britain's aircraft carriers

After the building of *Argus* (see 'Flight Deck') there then followed an uninterrupted line of steam turbine-powered aircraft carriers built for the Royal Navy, many serving with distinction in World War II. Probably the most famous is the third *Ark Royal* that was commissioned in 1938 and whose Swordfish torpedo bombers dealt the *coup de grace* to the German battleship *Bismarck* in May 1941. *Ark Royal* was sunk by a German U-boat off Gibraltar in November of that year. Her successor was launched at the Cammell Laird shipyard in Birkenhead in 1950 and was the world's first aircraft carrier to be fitted with an angled flight deck – she was decommissioned in 1978 and scrapped.

HMS Ark Royal

HMS *Hermes* became the last steam turbine-powered aircraft carrier to be commissioned for the Royal Navy – the smaller 'Invincible' Class jump-jet carriers and the newly-ordered 'Queen Elizabeth' Class are powered by gas turbines. *Hermes* had a long gestation period, being laid down in 1944, launched in 1953 and commissioned in 1959. She was refitted with a ski jump to operate Sea Harriers in 1980 and saw action during the Falklands War in 1982. She was decommissioned in 1985 but later refitted and sold to the Indian Navy where she was renamed *INS Viraat* in 1989. Despite her age she has been constantly modernised and is the oldest aircraft carrier in service in the world.

The Last Battleship
HMS Vanguard – *the end of a long-line*

By World War II the days of the mighty battleship were already numbered and by the end of the war air power had become supreme. The last in a long line of Royal Navy steam-powered battleships stretching back to the wooden, coal-fired HMS *Agamemnon* of 1852, HMS *Vanguard* was launched in 1944 and on commissioning in 1946 was the last battleship to be built in the world. Weighing 51,420 tons, with a speed of 30 knots and a complement of 1,500 crew, she was powered by four Parsons' steam turbines and was also the largest and fastest of the Royal Navy's capital ships ever.

Strangely, HMS *Vanguard* was mounted with four pairs of World War I vintage 15in guns which had been kept in store since the 1920s. They had previously been fitted to the battle cruisers *Glorious* and *Courageous* but were removed when these two ships were converted to aircraft carriers. HMS *Vanguard* never

HMS Vanguard

saw action as she was completed just after the end of World War II. She was decommissioned in 1960 and subsequently scrapped.

BULLEID'S 'LEADER' LOCOMOTIVES

Conversely, one of the strangest and least successful railway locomotives ever built in Britain was designed by one of its most innovative and successful locomotive engineers. Oliver Bulleid (see 'The Great Innovator') designed the slab-sided 'Leader' Class, which was conceived towards the end of World War II as a replacement for ageing tank locomotives. With a wheel arrangement of 0-6-0+0-6-0, it featured two six-wheeled steam-driving bogies, an offset boiler placed inside a box-like body, and duplicated driving cabs at each end linked by a corridor with a separate centrally-located cab for the firebox and fireman. Although five 'Leaders' were ordered only one, No. 36001, was completed, entering trials from Brighton Works in 1949. The locomotive soon proved to be poorly designed and highly unreliable and work was stopped on its four sister engines. Despite numerous modifications and improvements the prototype locomotive made its last run in 1950 and British Railways cancelled the project in the following year.

The End of the Line
Henry George Ivatt and Nationalisation

The son of H. A. Ivatt (see 'King of the Klondykes'), Henry George Ivatt was born in Dublin in 1886 when his father was the Chief Mechanical Officer of the Great Southern & Western Railway. Following a private education, he was apprenticed at the London & North Western Railway's Crewe Works in 1904 and by 1910 had gained experience as a draughtsman, assistant shed foreman and an assistant superintendent of outdoor machinery. Following war service in France, Ivatt was appointed Assistant Locomotive Superintendent of the North Staffordshire Railway, the company becoming part of the London, Midland & Scottish Railway in 1923.

Moving to the LMS's Derby Works in 1928, Ivatt was subsequently appointed as Locomotive Works Superintendent three years later. From 1932 to 1937 he was Divisional Mechanical Engineer for the LMS in Scotland before returning to Derby where he became the chief assistant to the CME, William Stanier (see 'A Swindon Man at Crewe'). Stanier retired in 1944 and was replaced by Charles Fairburn but his sudden death a year later catapulted Ivatt into the top position of Chief Mechanical Engineer of the LMS. Following Nationalisation in 1948, Ivatt stayed on as CME of the London Midland Region before retiring from BR in 1951. There then followed a spell as Director of Brush Traction in Loughborough before he finally severed his links with the railway industry in 1964. He died in 1976 at the age of 90 years.

Ivatt's output as a locomotive designer at the LMS and BR is fairly impressive and under his watch the last two 'Coronation' 'Pacifics' (Nos. 46256 and 46257) were built along with the first two main line diesel-electric prototypes (Nos. 10000 and 10001). The three new classes of steam loco designed by Ivatt were: 130 of the Class 2 2-6-2T; 128 of the Class 2 2-6-0 (known as 'Mickey Mouse'); and 162 of the Class 4 2-6-0 (known as 'Doodlebugs').

Doncaster's Last Great Man
Arthur Peppercorn

Arthur Peppercorn was born in Leominster, Herefordshire, in 1889 and began his railway career as an apprentice at the Great Northern Railway at Doncaster in 1905. He gained further experience as an Assistant Locomotive Superintendent at several GNR engine sheds before serving with the Royal Engineers during World War I.

Returning to Doncaster, he became Carriage & Wagon Works Manager for the newly-formed London & North Eastern Railway in 1923 and over the succeeding years held important posts at York, Stratford and Darlington and Doncaster before becoming Assistant Chief Mechanical Engineer under Edward Thompson (see 'A Mixed Bag') in

'A1' Class

1945. Thompson retired in 1946 and Peppercorn was promoted to Chief Mechanical Engineer, a position he held through Nationalisation when he was CME of the Eastern & North Eastern Regions of BR. He retired in 1949 and died in 1951.

Peppercorn is best known for his 'A1' Class 4-6-2s of which 49 were built between 1948 and 1949. Although renowned for their reliability they had an extremely short working life with the last example being withdrawn in 1966. His 'K1' 2-6-0s, of which 70 were built between 1949 and 1950, were also a successful, but short-lived, design.

High Standards
The 999 locomotives of Robert Arthur Riddles

Robert Riddles was born in 1892 (his place of birth is not recorded) and was a premium apprentice with the London & North Western Railway at Crewe Works between 1909 and 1913. During World War I he was seriously injured while serving with the Royal Engineers in France. Returning to Crewe Works after the war, he went on to become Head of the Production Department and was mainly responsible for the re-organisation of the LMS Works between 1925 and 1927. He repeated this task at Derby before being appointed in 1933 as Chief Assistant to the Chief Mechanical Officer, William Stanier (see 'A Swindon Man at Crewe'). During this period Riddles was closely involved in the design and building of the 'Coronation' 'Pacifics', was present on the record-breaking 114mph run of the 'Coronation' in 1937, and accompanied the locomotive's tour of North America in 1939.

Evening Star

During World War II, Riddles became Director of Transportation Equipment at the Ministry of Supply, designing the WD 'Austerity' 2-8-0 and 2-10-0 locos. He returned to the LMS in 1943 but was passed over in favour of George Ivatt (see 'The End of the Line') as CME following Charles Fairburn's sudden death. However, he was appointed Chief Mechanical & Electrical Engineer of the newly formed Railway Executive in 1947 and went on to design three classes of 4-6-2 ('Britannia', 'Clan' and 'Duke of Gloucester'), two classes of 4-6-0, three classes of 2-6-0, two classes of 2-6-2T, one class of 2-6-4T and one class of 2-10-0. Probably the most successful of his designs, 251 of the latter type were built and in 1960, No. 92220 *Evening Star* becoming the last steam locomotive to be built for British Railways. A job well done, with 999 of his standard locomotives eventually built, Riddles retired in 1953 and died in 1983.

ELIZABETHAN EXPRESS

Elizabethan Express *was a classic 20-minute film made by British Transport Films in 1954 to document the summer-only, non-stop 'Elizabethan' train between King's Cross and Edinburgh. The star of the film is 'A4' Class 4-6-2 No. 60017* Silver Fox *which was built at Doncaster in 1935 for use on the newly-inaugurated 'Silver Jubilee' express. She was withdrawn in 1963 and scrapped.*

You Thought It Was All Over…!
21st-century steam locomotives

By 1960 steam was on its last legs in Britain. Diesel-electric and gas turbines had long since replaced steam propulsion in ships, industrial steam and steam road vehicles were but a distant memory, and the last steam railway locomotive had just been built – within eight years these had totally disappeared from Britain's state-owned standard gauge railways.

However, the transformation over 50 years later is amazing. Not only did the new-build Class 'A1' 4-6-2 *Tornado* make its first appearance in 2009 along with the GWR steam railmotor in 2011, but also five more new-build locomotives are currently under construction. In the next few years we will be witnessing the rebirth of a BR Standard 'Clan' Class 4-6-2, GWR 'Grange', 'Saint' and 'County' Class 4-6-0s and an LMS un-rebuilt 'Patriot' Class 4-6-0. Apparently steam is not dead!

Steaming Across Britain
Where to see historic steam locomotives

The following is a small selection of heritage railways and museums where preserved steam locomotives can be seen in action.

BO'NESS & KINNEIL RAILWAY
**Bo'ness Station, Union Street, No'ness, West Lothian EH51 9AQ
Tel: 01506 822298
Website: www.srps.org.uk/railway**

BLUEBELL RAILWAY
**Sheffield Park Station, East Sussex TN22 3QL
Tel: 01825 720800
Website: www.bluebell-railway.com**

DARTMOUTH STEAM RAILWAY
**Queens Park Station, Torbay Road, Paignton, Devon TQ4 6AF
Tel: 01803 555 872
Website: www.dartmouthrailriver. co.uk**

EAST LANCASHIRE RAILWAY
**Bolton Street Station, Bury, Greater Manchester BL9 0EY
Tel: 0161 764 7790
Website: www.eastlancsrailway. org.uk**

FFESTINIOG & WELSH HIGHLAND RAILWAY
**Harbour Station, Porthmadog, Gwynedd LL49 9NF
Tel: 01766 516000
Website: www.festrail.co.uk**

GREAT CENTRAL RAILWAY
**Great Central Road, Loughborough, Leicestershire LE11 1RW
Tel: 01509 230 726
Website: www.gcrailway.co.uk**

ISLE OF WIGHT STEAM RAILWAY
**The Railway Station, Havenstreet, Ryde PO33 4DS
Tel: 01983 882 204
Website: www.iwsteamrailway.co.uk**

LLANGOLLEN RAILWAY
The Station, Abbey Road,
Llangollen, Denbighshire LL20 8SN
Tel: 01978 860979
Website: www.llangollen-railway.co.uk

MID-HANTS RAILWAY
Station Road, Alresford,
Hampshire SO24 9JG
Tel: 01962 733 810
Website: www.watercressline.co.uk

NATIONAL RAILWAY MUSEUM
Leeman Road, York YO26 4XJ
Tel: 08448 153139
Website: www.nrm.org.uk

NATIONAL RAILWAY MUSEUM
'Locomotion', Shildon, Co.
Durham DL4 1PQ
Tel: 01388 777999
Website: www.nrm.org.uk

NORTH NORFOLK RAILWAY
Sheringham Station, Sheringham,
Norfolk NR26 8RA
Tel: 01263 820800
Website: www.nnrailway.co.uk

NORTH YORKSHIRE MOORS RAILWAY
12 Park Street, Pickering, North
Yorkshire YO18 7AJ
Tel: 01751 472508
Website: www.nymr.co.uk

RIVERSIDE MUSEUM
100 Pointhouse Place, Glasgow
G3 8RS
Tel: 0141 287 2660
Website: www.glasgowlife.org.
uk/museums

SEVERN VALLEY RAILWAY
The Railway Station, Bewdley,
Worcestershire DY12 1BG
Tel: 01299 403 816
Website: www.svr.co.uk

SOUTH DEVON RAILWAY
The Station, Dart Bridge Road,
Buckfastleigh, Devon TQ11 0DZ
Tel: 013646 42338
Website: www.southdevonrailway.
co.uk

SWANAGE RAILWAY
Railway Station Approach,
Swanage, Dorset BH19 1HB
Tel: 01929 425 800
Website: www.swanagerailway.co.uk

WEST SOMERSET RAILWAY
The Railway Station, Minehead,
Somerset TA24 5BG
Tel: 01643 704996
Website: www.west-somerset-
railway.co.uk

MORE AMAZING TITLES

LOVED THIS BOOK?

Tell us what you think and you could win another fantastic book
from David & Charles in our monthly prize draw.
www.lovethisbook.co.uk

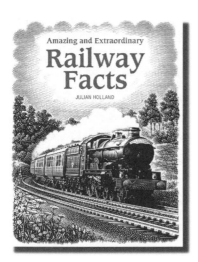

**AMAZING & EXTRAORDINARY
RAILWAY FACTS**
JULIAN HOLLAND
ISBN: 978-0-7153-2582-7

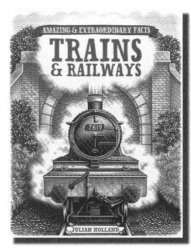

**AMAZING & EXTRAORDINARY
FACTS: TRAINS & RAILWAYS**
JULIAN HOLLAND
ISBN: 978-0-7153-3911-4

INDEX

A DAVID & CHARLES BOOK
© F&W Media International, Ltd 2012

David & Charles is an imprint of F&W Media International, Ltd
Brunel House, Forde Close, Newton Abbot, TQ12 4PU, UK

F&W Media International, Ltd is a subsidiary of F+W Media, Inc.,
10151 Carver Road, Cincinnati OH45242, USA

Text and designs copyright © Julian Holland 2012
Layout © David & Charles 2012

First published in the UK in 2012
Digital edition published in 2012

Layout of digital editions may vary depending on reader hardware and display settings.

Julian Holland has asserted the right to be identified as author of this work in accordance
with the Copyright, Designs and Patents Act, 1988.

A catalogue record for this book is available from the British Library.

ISBN-13: 978-1-4463-0193-7 Hardback
ISBN-10: 1-4463-0193-1Hardback

ISBN-13: 978-1-4463-5619-7 e-pub
ISBN-10: 1-4463-5619-1 e-pub

ISBN-13: 978-1-4463-5618-0 PDF
ISBN-10: 1-4463-5618-3 PDF

10 9 8 7 6 5 4 3 2 1

Acquisitions Editor: Neil Baber
Assistant Editor: Hannah Kelly
Project Editor: Freya Dangerfield
Design Manager: Sarah Clark
Production Manager: Beverley Richardson

Hardback edition printed in Finland by Bookwell Oy for:
F&W Media International, Ltd
Brunel House, Forde Close, Newton Abbot, TQ12 4PU, UK

F+W Media publishes high quality books on a wide range of subjects.
For more great book ideas visit: www.fwmedia.co.uk

Picture Credits: p26 © Chris McKenna, p31 © Richard Rogerson, p55, 69 © Chris Allen,
p56 © Ruth AS, p57 © Lord Price, p61 © Adrian Cieslak, p62 © Phil Parker, p71, 88, 100,
104, 133 © Ben Brooksbank, p74, 76 © Dr Neil Clifton, p110 © Aquitania, p117 © John
Griffiths, p119 © Ben Vicar, p123 © Boris Lux, p127 © David Ingham, p130 © Roland
Godefroy, p131 © David Bailey, p137 © Roger Cornfoot